Engineering Materials and Processes

Series Editor

Other titles published in this series

Fusion Bonding of Polymer Composites
C. Ageorges and L. Ye

Composite Materials
D.D.L. Chung

Titanium
G. Lütjering and J.C. Williams

Corrosion of Metals
H. Kaesche

Corrosion and Protection
E. Bardal

Intelligent Macromolecules for Smart Devices
L. Dai

Microstructure of Steels and Cast Irons
M. Durand-Charre

Phase Diagrams and Heterogeneous Equilibria
B. Predel, M. Hoch and M. Pool

Computational Mechanics of Composite Materials
M. Kamiński

Gallium Nitride Processing for Electronics, Sensors and Spintronics
S.J. Pearton, C.R. Abernathy and F. Ren

Materials for Information Technology
E. Zschech, C. Whelan and T. Mikolajick

Fuel Cell Technology
N. Sammes

Casting: An Analytical Approach
A. Reikher and M.R. Barkhudarov

Computational Quantum Mechanics for Materials Engineers
L. Vitos

Modelling of Powder Die Compaction
P.R. Brewin, O. Coube, P. Doremus and J.H. Tweed

Silver Metallization
D. Adams, T.L. Alford and J.W. Mayer

Microbiologically Influenced Corrosion
R. Javaherdashti

Modeling of Metal Forming and Machining Processes
P.M. Dixit and U.S. Dixit

Electromechanical Properties in Composites Based on Ferroelectrics
V.Yu. Topolov and C.R. Bowen

Charged Semiconductor Defects
Edmund G. Seebauer and Meredith C. Kratzer

Modelling Stochastic Fibrous Materials with Mathematica®
William W. Sampson

Ferroelectrics in Microwave Devices, Circuits and Systems
Spartak Gevorgian

Vladimir Kochergin • Helmut Föll

Porous Semiconductors

Optical Properties and Applications

Vladimir Kochergin, PhD
Luna Innovations, Inc.
3157 State Street
Blacksburg VA 24060
USA
vkotcherguine@msn.com

Helmut Föll, PhD
Universität Kiel
Technische Fakultät
LS Allgemeine Materialwissenschaft
Kaiserstr. 2
24143 Kiel
Germany
hf@tf.uni-kiel.de

ISSN 1619-0181
ISBN 978-1-4471-2676-8 e-ISBN 978-1-84882-578-9
DOI 10.1007/978-1-84882-578-9

British Library Cataloguing in Publication Data
A catalogue record for this book is available from the British Library

Softcover reprint of the hardcover 1st edition 2009

Cover design: eStudioCalamar, Figueres/Berlin

Printed on acid-free paper

Springer is part of Springer Science+Business Media (www.springer.com)

Foreword and Acknowledgements

Foreword

This book is the result of many years of research into porous semiconductors and their optical behavior and the culmination of an intense period of cooperation between the authors. The book is based on a substantial number of papers and other publications that the authors, together with a host of co-authors, have published since about 2004. It also includes of course the results from many other groups.

In addition to providing a coherent theoretical and practical overview of many aspects of porous semiconductors with respect to their optical properties and uses, the book contains a number of new topics and clarifications.

Many of the results given come from a cooperation of V. Kochergin, then mostly at Lake Shore Cryotronics, Inc., Columbus, OH, USA, with H. Föll, then and now holder of the Materials Science chair in the Institute of Materials Science, Faculty of Engineering, Christian-Albrechts-University of Kiel, Germany. The cooperation started when Marc Christophersen stayed for some time with Lake Shore after he obtained his PhD degree in Kiel. In 2005 and 2007 V. Kochergin spent some time in Kiel, and it is during these visits that the idea was born of putting together the materials related to optical properties and applications of various porous semiconductors in a coherent form. It took over two years, though, to summarize all the material and to augment it with the then missing experimental and theoretical parts.

Acknowledgements

The material of this book was prepared with assistance of many German and American colleagues of the authors. We would like to thank Dr. Marc Christophersen, Dr. Jürgen Carstensen, Dr. Sergiu Langa, Dr. Saman

Dharmatilleke, Dr. Kimberly Pollard, Mr. Mahavi Sanghavi, Mr. Ralph Orban, Mr. Russell Goose, Mr. Regis Toomey, and Mr. Jeff Hardman for help with sample preparation and characterization. We are also thankful for Prof. Ivan Avrutsky, Prof. Ulrich Gösele, Dr. Philip R. Swinehart, Prof. Ion Tiginyanu, Prof. Ralf B. Wehrspohn, and Dr. Vladimir Zaporojtchenko for fruitful discussions, contributions and advice. The help of Dr. Andreas Langner and Prof. Paul Bergstrom, who supplied pictures of their work, is gratefully acknowledged.

We are especially indebted to Ms. Katrin Brandenburg, who was of invaluable help with finalizing and editing the manuscript, and whose competence and patience we gratefully acknowledge.

One of the authors (VK) is especially indebted to Lake Shore Cryotronics, Inc. management for giving the financial support during the research and development of porous silicon optical components, as well as for permission to publish some of the material in this book. He would also like to thank personally Mr. Michael Swartz (COE/President), Dr. Philip R. Swinehart (VP R&D) and Dr. William McGovern (senior engineer and later department head) for generous access to the research material even after the author departed from Lake Shore. The author is also thankful for Dr. William McGovern for sharing the material used in a number of illustrations in this book that William collected after the author's departure from Lake Shore. V. Kochergin would further like to thank the late Dr. Volker Lehmann for sharing samples of macroporous silicon, and Dr. Ofer Sneh (Sundew Technologies, Inc.) for assistance with ALD coating of the macroporous silicon samples.

The authors acknowledge DFG, NASA, NSF, and MDA programs and the responsible program managers for the financial support of the research and development.

Last but not least the authors would like to acknowledge the direct and indirect support they received from their families while working on this book. Vladimir Kochergin would like to thank his wife Elena, his parents Eugene and Elena, and his children (Eugene, Katherine, and Anna); Helmut Föll thanks his wife Sara and his children (Daniel, Julia, and Alexandra). All of them would have preferred to have more time with the authors to having this book but gracefully accepted some sacrifice for the advancement of science.

Contents

1

Introduction

1.1 Pores in Silicon and Optics

Microporous Si produced by anodic dissolution of Si in HF bearing electrolytes had been known for quite some time; the first reports go back to Uhlir in 1956 [1] who, however, did not have the means to recognize the microporous nature of the "films" he produced by electrochemical dissolution. Later, after these films were recognized for what they were, microporous Si enjoyed some brief popularity in the 80ties since it was envisioned as the essential ingredient for integrated "FIPMOS" transistors; the acronym stands for "Full Isolation by Porous Oxidized Silicon". FIPMOS tried to exploit two outstanding properties of microporous Si: it could be produced selectively only from structured p-type areas in an n-type Si environment, and it could be oxidized much faster than bulk Si. Despite a conceptually attractive process, FIPMOS never made it into a mainstream Si technology, and one of the reasons for this can be found in the many problems encountered with "taming" the anodic dissolution of large wafers to an extent where the process would no longer count among the black arts but could be used with the tight tolerances, reliability and repeatability required by production technologies.

After that interesting but undistinguished history microporous Silicon was catapulted into prominence almost overnight when Canham [2] demonstrated that properly produced microporous Si layers, i.e. porous structures with dimensions (diameter and distance between pores) of the pores in the nanometer region, exhibited strong photoluminescence. Canham attributed this surprising finding to the "quantum wire" nature of the Si that was left between the pores. Lehmann and Gösele had proposed the same hypothesis a few months earlier [3] and used it for a first explanation concerning the so far unclear formation mechanism of microporous Silicon. Porous Silicon and optical properties, it appeared, made an irresistible combination for many engineers and scientists.

Alas, the history of luminescent microporous Si follows Oscar Wilde's (here slightly modified) dictum: "There are two tragedies in life: not to get your heart's desire and to get it". Silicon in its porous form finally could produce light if a proper junction to the porous layer was made, and thus met the heart's desire of all and sundry from the Si community. But what is light emitting Si really good for? Besides the fact that other semiconductors could do the job far better or cheaper, it was (and is) hard if not impossible to keep the emission from some light-emitting porous Si device stable over time.

Meanwhile, in 1990, Lehmann and Föll demonstrated another kind of pore in n-type silicon: Lithographically patterned macropore arrays with well-developed straight cylindrical pores and very large aspect ratios [4]. The essential "trick" for producing this kind of pores was to illuminate the backside of the substrate in order to produce the holes necessary for a substantial dissolution rate of n-type Si. The model proposed by Lehmann and Föll ascertained that the space charge region curving around the tip of growing pores would capture the holes diffusing in from the backside and thus induce and stabilize macropore growth.

In 1996 Lehmann and Grüning [5,6] showed that suitable macropore arrays constituted a two-dimensional photonic crystal (PC) and thus linked porous Si once more with a new branch of optics that caused considerable excitement after E. Yablonovitch and S. John established this branch of "photonics" [7,8]. Macropore arrays soon became popular in connection with photonic crystals; modulating the pore diameter with pore depth could produce even three-dimensional structures [9,10]. However, after more than 10 years of extensive R&D efforts, products like, e.g., sensors based on porous Si PCs, are still far from commercialization. Note that the direction of light propagation in a two-dimensional photonic crystal is perpendicular to the pore direction.

Once more, Lehmann was the first who linked pores in Si to another branch of optics by considering light propagation in the direction of pores (after he produced a macroporous membrane by polishing off the unetched region of the substrate) [11]. It is an interesting little puzzle to consider off-hand how light would pass down a "light sieve", i.e. a regular array of many holes with diameters and lattice constants in the μm range and lengths of several 100 μm. What might come to mind is that "big" photons would get stuck whereas "thin" photons, i.e. photons with wavelength smaller than the pore diameter, would be able to pass through the sieve, forming a Bragg diffraction pattern upon exiting the membrane. A first quantitative analysis done by Avrutsky and Kochergin [12] proved that this somewhat naïve concept is not too far off the truth: macroporous membranes are indeed high-pass filters with the capability to transmit, for example, ultraviolet light while completely blocking larger wavelengths. While this first analysis was somewhat too simple, the inherent basic concept proved to be sound: Light can be processed in many ways by porous materials by adjusting the geometry of the pores in a substrate. In other words: Silicon or other semiconductors with suitable pores constitute an optical metamaterial with adjustable optical properties that can be scaled from the far infrared (IR) to the extreme ultraviolet (UV) "simply" by scaling the geometry of the pores. This is particularly true for light propagation in pore direction; the approach thus goes beyond photonic crystals.

1.2 Organization of this Book

While a quick thumbing through the book might indicate that a more theoretical approach was taken with respect to the topic “optical properties of porous semiconductors”, this is actually not the case. Most of the various optical devices presented in detail in what follows have actually been made and tested. However, the results of many wasted weeks of experimental work in the laboratory, where it was found once more that pore etching in Si, while nowadays based on an extensive set of knowledge and experimental ken, does still have a somewhat mysterious component (cf. e.g. the newest findings of the group of one of the authors (hf) [13]), does not need much space to report. Nevertheless, the book contains a large number of experimental results that at the minimum provide a proof-of-principle and at best indicate the possibility of starting pilot production.
The task of producing porous metamaterials with properties that can be adjusted and maintained with the kind of precision that is so typical for optical components will still be a challenging enterprise when this book appears in print. While the present state-of-the-art of pore etching may not yet be up to production standards meeting the necessary precision for most of the optical components discussed in this book, the authors are convinced that electrochemical pore etching, given enough attention, can be developed to a precision tool just as, e.g., plasma etching some 20 years ago.

The second chapter therefore introduces not only the basics of the electrochemistry on pore etching in Si in a way that does not require background knowledge of electrochemistry but moves on to rather recent developments with regard to a better understanding of pore etching and first attempts at in-situ control. While there is still no reliable theory that relates pore properties to etching parameters in detail, there are good reasons to believe that on-going progress in pore etching will soon reach a state where even the more complex optical devices discussed in this book can be made with sufficient precision.

The third chapter deals with the “subwavelength” mode of light propagation through porous semiconductors, i.e. with situations where the geometrical parameters of the pore are much smaller than the wavelength of the light. In this case the porous semiconductors can be considered as optically homogeneous and described by an effective index of refraction. This is nothing new and a review of effective media theories will be given first in this chapter. Since certain pore geometries break the symmetry of the underlying (cubic) crystal, optical anisotropy is encountered and the index of refraction becomes a tensor. Going one step beyond this kind of (already known) anisotropy (with effects like double diffraction) by now breaking the symmetry of the pore array, leads to predictions of new and unusual properties of porous materials, expressed in tensors of a kind never encountered in natural materials. These kinds of porous semiconductors may truly be called metamaterials.

Chapter 4 discusses the “light sieves” mentioned above and begins where the old approach for light transmission through porous membranes ended. The pore with its surroundings, being the opposite of a standard waveguide with the high index material in the inside and the low index material as cladding, has still light guiding properties if the pore is much longer than the wavelength. It can be treated as a "leaky waveguide" and this approach is developed in some details since it forms the base for some of the following chapters.

Chapter 5 deals with an effect quite simple but taken into account rather late: The Si between straight pores in a periodic array forms a conventional if geometrically complex waveguide. It will transmit light with energies below the band gap with many of the characteristics of regular wave-guides. While this is an unwelcome effect for, e.g., UV filters, it offers possibilities of its own. A detailed description of this mode of light transmission through porous semiconductors is therefore given.

Chapter 6 introduces the first true device based on mesoporous Si. The basic idea is to use the possibility of changing the effective index of refraction as a function of depth in a multitude of ways. This has been a playing field of many groups for more than 10 years and the state-of-the-art will be reviewed first. Some experimental efforts with a focus on filters for far IR applications will be presented, and particular emphasize is given to the long term stability of these filter types, the major unsolved problem in this field with respect to commercialization.

Chapter 7 introduces long wave pass filters, i.e. filters that block transmission for wavelengths above a certain critical size. In this case the light scattering properties of porous layers is employed. This is, perhaps, the application that makes the least demands on pore parameters and it is thus no surprise that it is the only optical device based on porous Si that is actually on the market since 2005.

Chapter 8 takes up in earnest the topic of UV filters discussed before and dealt with on a theoretical base in Chapter 4. For this particular application, porous Si offers a number of advantages in comparison to established technologies that must rely on cumbersome total reflection properties of mirrors. In this chapter the coating of pore walls with suitable dielectrics or metals in order to increase the performance is first introduced; it will be used throughout what follows.

Chapter 8 leads directly up to Chapter 9 where polarization components for the UV range are considered. Again, existing technologies for this kind of optical component, while complex, do not result in very good devices, and porous Si could provide for a better performance. All it takes are pores with elongated cross sections, and a proof of principle is provided. However, good polarizers based on this concept need yet to be made. The progress made in pore etching since the first efforts along this line make it likely that a new endeavor in this respect would be fruitful.

Chapter 10 discusses retroreflection suppression plates, an application that comes straight from military technology but will find civilian uses too. Like the long wave pass filters introduced in Chapter 7, it should not be very difficult to make.

In contrast, Chapter 11 introduces the most difficult to make but possibly also the most enticing of all optical devices based on porous semiconductors: omnidirectional filters for the IR, i.e. band pass filters with characteristics that do not depend on the angle of incidence. This kind of filter is unique to porous materials but demands precise control of the pore geometry in three dimensions. In particular, it needs pores with diameters that change periodically with depth, and that is the reason why pore diameter modulations have dealt with in some detail in Chapter 2. While a proof of principle has not yet been possible there is little doubt that with the improved pore etching technologies that emerged in the meantime, good results could be obtained.

Chapter 12 carries optical engineering with porous semiconductors to the field of bio sensing. It serves to demonstrate that optics and porous semiconductors may be combined for new functionalities transcending optical components in many new an innovative ways. Since sensing applications of porous silicon have been thoroughly reviewed before, the chapter provides mostly a brief overview of the state of the art, particularly in the area of optical bio-chemical sensors.

1.3 References

[1] Uhlir A, (1956), Electrolytic shaping of germanium and silicon. Bell System Tech. J. 35:333–347.

[2] Canham LT, (1990), Si Quantum Wire Array Fabrication by Electrochemical and Chemical Dissolution. Appl. Phys. Lett.57:1046–1048.

[3] Lehmann V, Gösele U, (1991), Porous Silicon Formation: A Quantum Wire Effect. Appl. Phys. Lett. 58:856–858.

[4] Lehmann V, Föll H, (1990) Formation mechanism and properties of electrochemically etched trenches in n-type silicon. J. Electrochem. Soc. 137:653–657.

[5] Grüning U, Lehmann V, Ottow S, Busch K, (1996) Macroporous silicon with a complete two-dimensional photonic band gap centered at 5 um. Appl. Phys. Lett. 68:747–749.

[6] Grüning U, Lehmann V , Eberl U, (1996) Photonische Bandstruktur in Makroporösem Silizium, Phys. Bl. 52: 661.

[7] Yablonovitch E, (1987), Phys. Rev. Lett. 58:2059–2062.

[8] John S, (1987), Phys. Rev. Lett. 58:2486–2489.

[9] Wehrspohn RB, Schilling J, Choi J, Luo Y, Matthias S, Schweizer SL, Müller F, Gösele U, Lölkes S, Langa S, Carstensen J, Föll H, (2004) Photonic Crystals, Wiley-VCH, Weinheim, 2004 p. 63.

[10] Matthias S, Müller F, Jamois C, Wehrspohn RB, Gösele U, (2004), Adv. Mater. 16:2166.

[11] Lehmann V, Stengl R, Reisinger H, Detemple R, Theiss W, (2001) Optical shortpass filters based on macroporous silicon. Appl. Phys. Lett. 78:589–591.

[12] Avrutsky I, Kochergin V, (2003) Filtering by leaky guided modes in macroporous silicon. Appl. Phys. Lett. 82: 3590–3592.

[13] Ossei-Wusu EK, Cojocaru A, Carstensen J, Leisner M, Föll H, (2008) Etching deep macropores in n-type silicon in short times, ECS Trans. 16:109.

2

Pore Etching Essentials

2.1 Scope of this Chapter

This book is primarily intended for the optical community, which cannot be expected to be very familiar with electrochemistry in general and with electrochemical pore etching in Silicon in particular. In the space available no serious attempt will be made to educate the average reader in this respect. Instead, basic features of pore etching will be discussed; as far as possible from the viewpoint of a physicist. For following this chapter thus neither detailed knowledge of electrochemistry is required, nor the usually unfamiliar terminology. This will lead to some imprecision and approximations in parts of this chapter, but that does not distract too much from what we will attempt here: to give a general idea of what can be done by pore etching in Si at present, how it is done, and where the limits are now and in the foreseeable future. Dyed-in-the-wool electrochemists may shudder at this but the approach is nevertheless justified because "serious" electrochemistry cannot yet give a coherent picture of pore etching in terms of a theory with some predictive power; what kind of approach is taken to elucidate some of the findings is such largely a matter of taste. For a more detailed overview and discussion of the proper electrochemistry the reader should consult the books of Lehmann [1] and Zhang [2], reviews like [3–9] and in particular the very recent review about macropores in Si [10] as well as the proceedings of the bi-annual Int. Conference on Porous Semiconductor Science and Technology (PSST) [11–14].

We will first look at some necessary definitions and then proceed to the bare necessities for electrochemical etching; on occasion in a more or less axiomatic way, i.e. by just giving some facts without any further reasoning. This is followed by a description of the typical set-up of a pore etching experiment with a relatively detailed discussion of the available parameter space. Next we discuss the "classical" case of macropore etching in lightly doped n-type Si under backside illumination in the context of the "space charge region model" including the so-called "Lehmann formula". This will develop into a critical discussion of the state-of-the-art based on rather recent findings with respect to macropore etching in n-type Si and a brief introduction and discussion of a more general model for pore formation in the context of the so-called "current burst model". In addition, a description of the just evolving in-situ monitoring technique employing multi-mode in-situ FFT impedance spectroscopy will be given. The chapter ends with a foray into macropore etching in lightly doped p-type Si, where a new view at this topic will be proposed, some remarks on meso- and micropore etching, and a very short look at pores in III-V semiconductors.

2.2 Some Basics and Definitions

2.2.1 The Essentials

Electrochemical pore etching in semiconductors involves running some current I through a semiconductor-electrolyte junction by connecting the back side of a semiconductor "**working electrode**" and an ideally featureless **counter electrode** to a **potentiostat** or **galvanostat** as schematically illustrated in Figure 2.1. That part of the semiconductor that is exposed to the electrolyte is called the front side here. An area A of the semiconductor is exposed to the electrolyte, and it is good practice to refer to the global current density $j_g = I / A$ instead of the rather meaningless external current I. In practice, the area A varies from a few mm^2 in laboratory experiments to ≈ 30.000 mm^2 for a standard 200 mm Si wafer; a typical current density of 50 mA/cm^2 thus translates into currents between ≈ (5–15.000) mA for the areas given. Some external voltage V, typically a few V but up to ≈ 100 V on occasion drives this current. Looking at the extremes it becomes already apparent at this point that the potentiostat/galvanostat could be a simple device sufficient for delivering a few 100 mA at < 10 V or so, or a big piece of hardware with a power consumption in the kW range, perfectly capable of electrocuting the experimentalist. It is useful at this point to note that the voltage drop across the immediate solid-liquid junction, i.e. the **potential** that drives the chemical reactions, is always small, like for a forwardly biased pn-junction; otherwise very large current densities would result. If the external voltage or potential is large, the major voltage drop must be somewhere else, e.g. in an oxide layer or in the space charge region in the semiconductor.

Let's now turn to silicon. The "classical" macropores in Si are obtained by **back side illumination** (bsi) of lightly doped n-type Si; usually with lithographically defined nuclei for the desired pore arrangement. We will not go into technical details like how to make lithographically defined structures here; this topic is either familiar or covered in many monographs and papers. Finding a mask that survives many hours of etching might be demanding, however; cf. [15–17]. The illumination intensity P produces a number of electron – hole pairs or injects a photocurrent density j_P at the backside simply given by the number of holes produced per cm^2 and second. The electrolyte is typically **aqueous** hydrofluoric acid (HF), possibly mixed with ethanol or some other ingredients, and with the HF concentration typically around 5 wt. %. While the remark about potentially deadly power supplies might have been superfluous to physicists, we must remark at this point that HF is extremely dangerous and must be handled with the utmost care!

We refer to the macropores obtained in this way by the self-explaining shorthand notation introduced in [3] as: n-Si-macro(aqu, bsi, litho) pores.

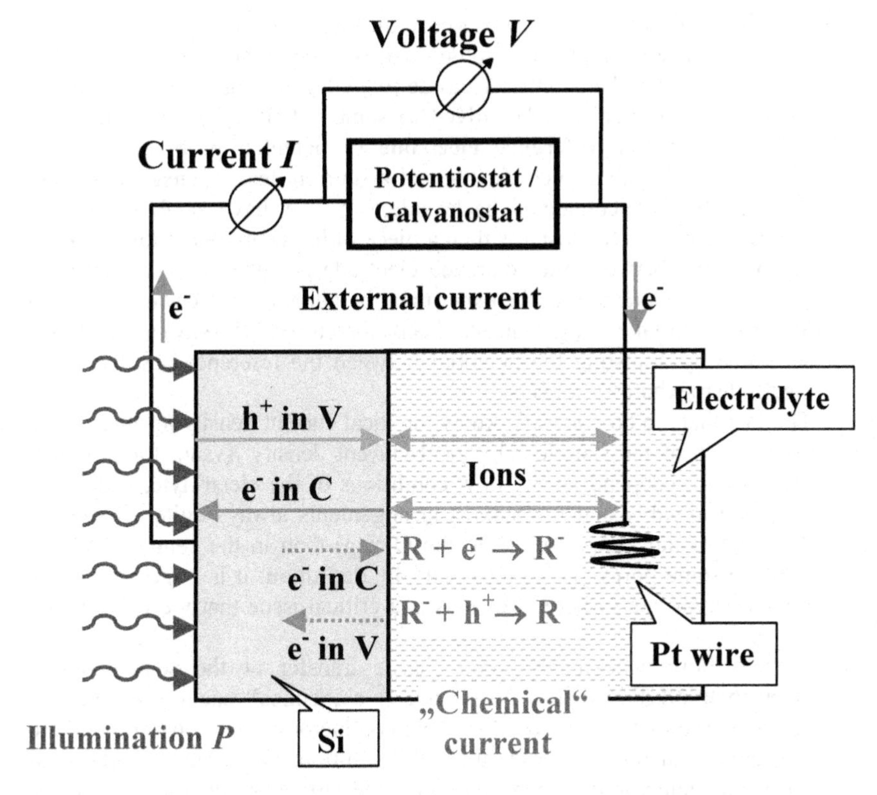

Figure 2.1. Principles of current flow upon anodically dissolving Si. Note the separation into external currents and "chemical" currents

Note that the external current is always carried by electrons outside of the electrolytical cell, but decomposes into electron and hole currents in the semiconductor and into an ionic current, consisting of possibly several ions with positive and negative charges. Moreover, besides the external current a "chemical" current may provide for some charge transfer across the semiconductor – electrolyte interface, reducing or oxidizing some chemical species contained in the electrolyte. This is simply due to the fact that the semiconductor – counter electrode system is always a battery with some build-in potential difference. Note also that an external current of $I = 0$ A does not mean that nothing happens. It simply could mean that the cathodic current (electrons flowing out of the semiconductor or holes flowing into the semiconductor) equals the anodic one. While the two partial currents may cancel each other, the chemical reactions tied to any current flow still will take place.

With the system described so far we can endeavour to characterize the solid – liquid junction by measuring its current (density) – voltage (*IV*) characteristics with the light intensity *P* (and other variables) as parameters just as one could do it for the Schottky junction we would have if we replace the electrolyte with a metal. There are, however, a few trivial and a few not so trivial problems associated with doing that experiment. On the trivial side we have:

1. The fact that we do not really know the potential at the semiconductor interface. There might be some sizeable potential drop across the electrolyte and at the counter electrode (which is not really featureless) and so on. This problem can be solved to some extent by introducing a so-called current-free **reference electrode** (a potential probe in physical terms) close to the semiconductor and a (current-less) **"sense" electrode** right a the semiconductor back side; i.e. we have a four terminal arrangement. A potentiostat is then a piece of hardware that allows to keep the potential between the reference electrode and the sense electrode at some pre-set value during the experiment by adjusting the current (density) to whatever it takes. A galvanostat keeps the current (density) at some pre-set value by adjusting the potential between the reference and the sense electrode to whatever it takes.
2. The fact that we do not really know the local current density $j(x, y)$ at some point (x, y) of the sample. The local current density $j(x, y)$, for example, may strongly depend on the flow conditions of the electrolyte, which can vary considerably and is for most arrangements always quite different at the edge of the sample (close to the O-ring) than in the centre. If we do produce some pores, $j(x, y)$ must vary by definition: it is large at the pore tip and small everywhere else. This is a critical issue that we will take up again.
3. Current flow necessitates some charge transfer at the semiconductor electrolyte interface that is always tied to a chemical reaction. For **anodic currents** (holes flowing out of the semiconductor or electrons flowing into the semiconductor) this reaction is Si dissolution. The surface condition of the semiconductor thus changes with time (for example its surface area increases dramatically if it becomes porous), again raising some questions as to the meaning of the local current density $j(x, y)$ with respect to the area.

Despite these valid points one can certainly always run an *IV* curve and make an attempt to interpret it. The expectation is, of course, that we will find some (Schottky) diode-like behavior since in a first approximation we can consider the electrolyte to behave like a metal. A strong **forward** current should flow if n-type Si is cathodically biased (minus pole at the Si) or if p-type Si is anodically biased because this allows easy transfer of the majority carriers to the electrolyte. The chemical reactions going with this are H_2 evolution on the cathodic side ($H^+ + e^- \rightarrow H$) and Si oxidation or dissolution on the anodic side ($Si + 4\ h^+ \rightarrow Si^{4+}$). If the polarity is changed, a small reverse current is to be expected in the dark and some photocurrent if the electrode is illuminated.

2.2.2 Basic Current-Voltage Characteristics

The paradigmatic potentiostatic (lightly doped) Si – aqueous HF electrolyte *IV* characteristics as shown in Figure 2.2 show this expected behavior to some extent but also many peculiar features not found in real (Schottky) diodes.

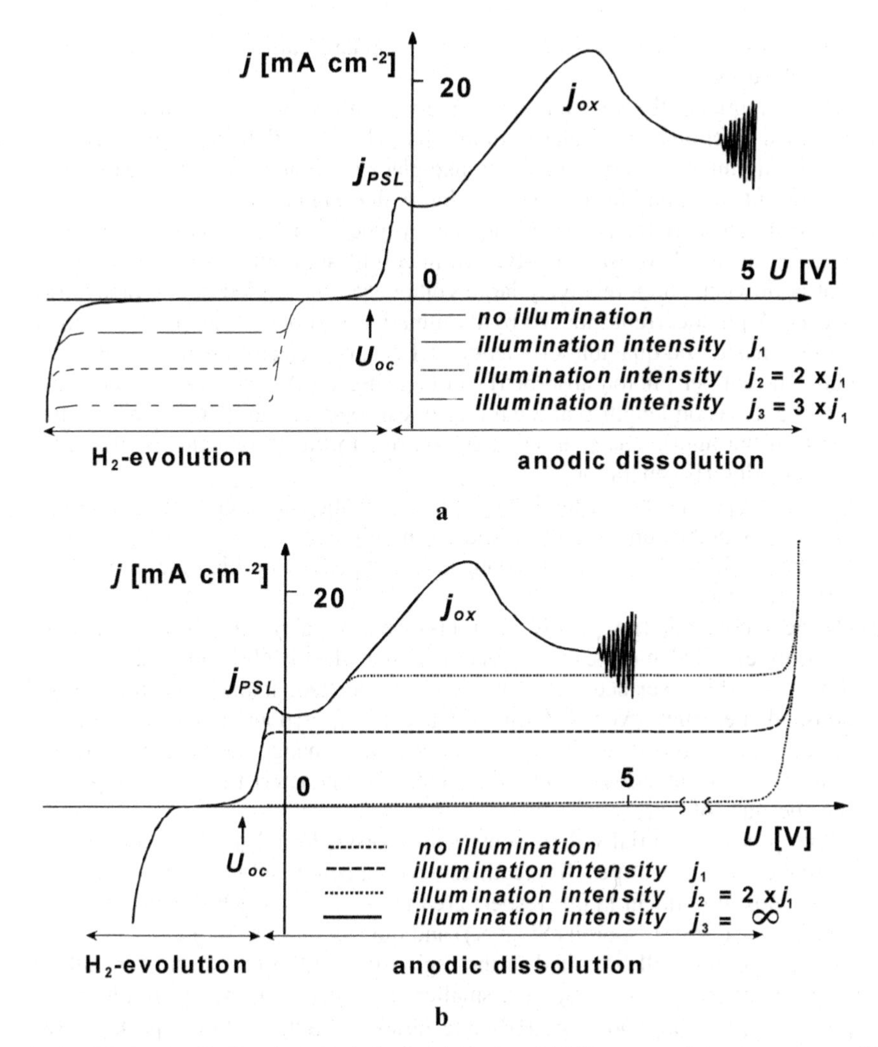

Figure 2.2. Basic Si *I*(*V*) characteristics in about 1 wt% HF, **a** aqu. electrolyte, p-type Si, **b** aqu. electrolyte, n-type Si This makes clear that we go beyond semiconductor junction behaviour and encounter "electrochemistry" at this point. Suffice it to say that the current density of both peaks increases about linearly with the HF concentrations as long as [HF] < 10 % or so

Discussing the *IV* curves shown in any detail would far exceed the scope of this chapter we will thus only enumerate some of the more pertinent features for this type of characteristics:

i) There is considerable "aberrant" behavior around 0 A, which is not found at $V = 0$ V! One reason for this is simply that the "battery" voltage V_{Bat} of the chemical battery formed by the "working electrode" Si, the counter electrode and the electrolyte must be added to the externally applied voltage. Since V_{Bat} depends on many parameters, (e.g. the counter electrode material) the voltage scale is usually referenced to "SCE", the standard calomel electrode as reference electrode.

More important, the reaction kinetics for small voltages is sensitive to the chemical reactions taking place. Simply put, H_2 generation and so on "needs" a certain minimum "overpotential" to take place at all and this is the reason why the currents are small in some voltage interval around $I = 0$ A.

ii) The cathodic part for p-type Si and for n-type Si behaves as expected for a reverse bias > 2 V or so. For p-type Si in equilibrium only a small reverse current is flowing, and a relatively large voltage can be applied before breakdown occurs. A photocurrent induced by illumination is proportional to the illumination intensity; and quantum efficiency of 1 (i.e. one electron contributing to the photocurrent per photon absorbed) can be achieved if conditions are favorable (i.e. the diffusion length L of the minority carriers is larger than the penetration depth of the light). The chemical reaction tied to the current flow is the rather uninteresting H_2 generation.

For n-type Si the junction is in forward conditions; the current grows exponentially with the applied voltage and essentially generates H_2.

The Si electrode remains unchanged in both cases; no dissolution or oxidation takes place.

iii) On the anodic side (i.e. positive currents) we always dissolve Si in HF bearing electrolytes. That means, quite generally, that the Si electrode changes with time – its shiny surface may get rough or facetted, it may become electro-polished (i.e. shiny even if it was rough before), or pores nucleate and grow, producing a porous layer. Whatever happens will change the experimental conditions, and we must expect that the anodic *IV* characteristics will change with time because of that.

iv) While p-type Si initially does show some forward-kind diode characteristics, there are a number of peculiar features never encountered with solid-state junctions. Most prominent in this respect are the two peaks: the PSL peak at current density j_{PSL} ("PSL (= porous Si layer)) and the "oxide peak" at j_{ox}.

v) Pore generation will be found in the region before the PSL peak; typically p-macro(aqu) pores at $j << j_{PSL}$ and smaller HF concentrations and p-micro(aqu) pores for $j <\approx j_{PSL}$ and large HF concentrations. Behind the j_{ox} peak electro-polishing is observed, always accompanied by self-induced current oscillations under potentiostatic conditions and voltage oscillations under galvanostatic conditions. This is the first example of a self-ordering phenomenon encountered at semiconductor electrochemistry; more details to this point can be found in [18] and in what follows.

vi) With n-type Si in the dark we find a reversely biased junction as expected. With illumination we now find a surprise: The quantum efficiency may be around 2 or even larger – especially for smaller illumination intensities and thus photocurrents.

If we use arbitrarily intense illumination and thus produce more holes than the electrolyte can process, the n-type *IV* characteristics is similar to that of p-type Si, as one would expect.

Pore generation will be found in most parts of the area in *IV*-space with currents below j_{PSL} and for all currents flowing because of electrical breakdown of the junction.

vii) If we now turn to the subject of pore etching more closely, it is imperative to note that for p-type Si any "working point" (i.e. initially defined external current and voltage pair (I, V)) for a pore etching experiment must lie on the *IV* curve of the system, while for n-type Si any working point below the limiting curve for intense illumination can be chosen by suitable illumination conditions. In other words, for p-type Si we have only **one** free (prime) parameter (voltage **or** current), whereas for n-type Si we have **two** (e.g. voltage V **and** current I – with the illumination intensity P then fixed).

viii) There is no theory or equation that describes the measured curves in some detail – in contrast to real pn-junctions or Schottky contacts. That also means that we cannot predict precisely how the basic *IV* characteristics change with the temperature T, the basic parameters of the Si (essentially doping concentration N_{dop}, and minority carrier life time τ or diffusion length $L = [D\tau]^{1/2}$ (D = diffusion coefficient of the minority carriers)) and the basic parameters of the electrolyte (essentially the HF concentration [HF] and the nature of the solvent for the HF (e.g. water H_2O)). Of course, a lot is known from general principles and from experiments; refer to [1,2] for data.

At this point it becomes necessary to point out expressively that the *IV* characteristics of both types of heavily doped Si (p^+ and n^+) may be quite different from the ones shown in Figure 2a and b and that the same is true if we resort to "organic" electrolytes; i.e. electrolytes where the always necessary HF is dissolved in organic solvents like acetonitrile, DMF, and DMSO instead of in water. Refer to [3,19,20] for details, here we only note that the strong quantitative changes in the *IV* characteristics under these conditions are mirrored in strong changes of the pores resulting from a suitable etching experiment under these conditions. Very much simplified it can be stated that etching of n^+ and p^+ Si under almost all conditions always results in mesopores, while organic electrolytes tend to produce mostly macropores.

2.3 Dissolution Mechanisms and Pore Formation

2.3.1 Basics

Pore formation implies localized dissolution of Si and thus is restricted to the anodic part of the *IV* characteristics. It also implies that the local current density $j(x, y, z, t)$ is a strong function of the coordinates (including the depth z in this case) and possibly the time t. The growth speed $v_{Pore} = \mathrm{d}z / \mathrm{d}t$ of some cylindrical pore with radius r is directly given by v_{Pore} = number of Si atoms dissolved in the volume $\pi r^2 \mathrm{d}z$ during differential time $\mathrm{d}t$, times the valence n of the dissolution process, times total amount of elementary charges passed in $\mathrm{d}t$, or

$$\rho_{at} \pi r^2 \,\mathrm{d}z = n \cdot j(x, y, z, t)\, \pi r^2 \cdot \mathrm{d}t \qquad (2.1a)$$

$$dz / dt = v_{Pore} = n \cdot j(x, y, z, t) / \rho_{at} \tag{2.1b}$$

with ρ_{at} = atomic density of Si = $5 \cdot 10^{22}$ cm^{-3} atoms/cm^3.

The **valence** n is a decisive parameter for pore etching, not least because it is one of the few quantities that can be easily measured ex-situ: just correlate the weight loss Δm after etching to the total number of charges $Q = \int I dt$ run through the interface and you have

$$n = (q / e) \cdot (\Delta m / m_{Si}) , \tag{2.2}$$

with m_{Si} = atomic mass of Si ($4.6638 \cdot 10^{-23}$ g).

Note that the term "valence" here has nothing to do with the term "valence band". It typically has values between $n = 2$ and $n = 4$ and is tied to the exact nature of the dissolution process. With respect to the *IV* characteristics as shown in Figure 2.2, the valence n changes from n ≈ 2 before the PSL peak to n ≈ 4 after the PSL peak. It is a known if not well understood fact that for standard n-macro(aqu, bsi) pores the valence needs to have a value of $n \approx 2.7$ if good pores are desired. This gives a first hint that pore etching in Si requires to take a closer look at the actual dissolution processes or the (electro)chemistry of the system.

Dissolution happens at the Si surface; i.e. Si atoms sitting at the "surface" of the sample will eventually end up in the liquid in some form of a (complexed) ion (in the end always $H_2^{2+} - SiF_6^{2-}$). If we look at a particular Si atom sitting in the crystal lattice close to the front side, its dissolution history at the most fundamental level is Si(bulk) → Si(interface) → Si(solution). Spelling out this simple sum reaction in more detail including the necessary chemical reactions quickly leads to > 50 detailed reaction paths, which we will not consider here at all; cf. e.g. [2], In this chapter we just focus on a more semiconductor based description that emphasizes the carrier flow across the interface. Including carrier flow, the gross reaction of removing one Si atom from the c atoms building up the crystal can always be described by the simple equation

$$[c\mathrm{Si}]_{cryst} + 4h^+ \Rightarrow [(c-1)\,\mathrm{Si}]_{cryst} + [\mathrm{Si}^{4+}]_{sol} , \tag{2.3}$$

(with h^+ denoting electronic holes in the Si) because in the grand total four bonds must be broken to release one Si atom from the solid. Within this description four holes concentrate on one Si atom, finally removing it from the crystal, and this process would be exclusively a **valence band process**. Holes move in the valence band to where they are needed; if the corresponding hole current needs to be converted into an electron current at the sample front and back side, electrons recombine with the holes, i.e. they are injected into the valence band. The four holes are thus consumed in the process by recombining with electrons provided by the "electrolyte" at the interface. The Si crystal is now four holes short of equilibrium and the holes consumed in the reaction must be replaced to keep the process going. This is what the current source does – it removes four electrons at the back side, which is equivalent to injecting four holes into the valence band, and transfers these electrons to some ions in the electrolyte at the counter electrode. Note that for this process four elementary charges in the form of electrons must flow through the external electric circuit in order to dissolve one Si atom – we have valence $n = 4$ by definition.

The dissolution process described by Equation 2.3 should be an easy process for p-type Si where plenty of holes are available and disturbances of the equilibrium conditions of the majority carriers near the interface will be relaxed in picoseconds, i.e. within the dielectric relaxation time, and on distances in the order of the Debye length, i.e. on the nm scale. In other words, the magnitude of the current flowing across the interface, and thus the reaction rate of the Si dissolution process, can be expected to be dominated from parameters other then the hole concentration in the Si.

In n-type Si the situation is completely different. Without an outside source of holes, only a few holes (present at an equilibrium concentration of $n_h = n_i^2 / n_e^- \approx n_i^2 / N_D$ (with n_i = intrinsic carrier concentration, N_D = doping concentration) are available for the reactions, and the maximum consumption rate (or current density j) might be reasonably expected to follow standard semiconductor junction theory. It thus should scale with the generation current for a reversely biased (asymmetric) junction, which for Si is given by [21]

$$j = (\mathrm{e} \cdot n_i^2 \cdot L) / N_D \cdot \tau + (\mathrm{e} \cdot n_i \cdot d_{SCR}) / \tau \approx (\mathrm{e} \cdot n_i \cdot d_{SCR}) / \tau \qquad (2.4)$$

with e = elementary charge, L = diffusion length (of the holes), τ = life time (of the holes), and d_{SCR} = width of space charge region (SCR). The first term describes contributions from generation in the bulk; the second term approximates generation in the space charge region (SCR). The approximation made is valid for semiconductors with relatively large band gaps like Si (but not, for example, for Ge), since in this case the reverse current characteristics are dominated by carrier generation in the SCR. If thermal generation in equilibrium is the only source of holes, the maximum anodic current would be close to zero (realistically some $\mu A/cm^2$). If non-equilibrium hole sources like illumination or local SCR breakdown (e.g. by avalanche or tunneling effects at high field strength) come into play, the anodic current and thus the chemical reaction rate at the interface could be much larger.

At this point the Si-electrolyte junction can be considered to be quite similar to an asymmetric pn-junction (with the electrolyte part corresponding to a highly doped side) or a Schottky junction; if we include illumination effects we essentially discuss a solar cell. For the current (density) – voltage characteristics (*IV* curve) we thus expect basic diode behavior, and that is what we get (up to a point) as shown in Figure 2.2. However, as pointed out already, the quantum efficiency in this case may be > 1, i.e. one photon causes more than one elementary charge to flow in the external circuit. In essence this demonstrates that Si dissolution could also be obtained – in principle – by a **conduction band process**, i.e. by **injecting electrons** in the conduction band according to

$$[c\mathrm{Si}]_{cryst} \Rightarrow [(c-1)\mathrm{Si}] + [\mathrm{Si}^{4+}]_{sol} + 4\mathrm{e}^- . \qquad (2.5)$$

In other words: Instead of picturing the transfer of a Si atom – or better Si^{4+} – from the crystal into the solution by "cutting" the four bonds by four holes, we equally well could just "rip out" an atom, leaving four electrons behind. Hole extraction or electron injection, as ever so often, produce equivalent results.

The two extreme processes involving only holes or only electrons electron as summarily given by Equations 2.3 and 2.4 are not observed, however. The truth, as ever so often, is a compromise combining both extremes according to

$$[N\mathrm{Si}]_{\mathrm{cryst}} + \lambda\, \mathrm{h}^{+} - (4-\lambda)\, \mathrm{e}^{-} \Rightarrow [(N-1)\, \mathrm{Si}]_{\mathrm{cryst}} + [\mathrm{Si}^{4+}]_{\mathrm{sol}} \,. \tag{2.6}$$

The parameter λ could have any value between 0 and 4, but is usually found to be > 1. The case $\lambda = 1$ can be most simply imagined as follows: A first hole cuts one bond of a surface atom (where two bonds where cut previously) while the remaining last bond might just "rip", leaving behind (= injecting) one electron. Supplying one hole, e.g. by the absorption of a photon, then leads to a current of two charges in the external circuit. In total, we still need four charges to dissolve one Si atom, so the valence of the process should still be $n = 4$.

No real chemistry so far, but now we must involve it to some unavoidable extent. To make a complicated story short, two very different dissolution processes coexist in Si: direct dissolution and dissolution via oxidation followed by oxide dissolution.

2.3.2 Direct Dissolution

Direct dissolution might proceed as described in [2] via

$$\mathrm{Si} + 6\, \mathrm{HF} + \lambda\, \mathrm{h}^{+} \Rightarrow \mathrm{H_2SiF_6} + 4\, \mathrm{H}^{+} + (4-\lambda)\, \mathrm{e}^{-} \,. \tag{2.7}$$

If the holes are produced by illumination, quantum efficiencies could have any value between 1 ($\lambda = 4$) to 4 ($\lambda = 1$). In other words, the photocurrent measured (hole current plus current from injected electrons) is up to 4 times larger than what would be expected for the photo generated holes only. While the valence n of the process in Equation 2.7 is still $n = 4$, direct dissolution is often called **divalent dissolution**, i.e. $n = 2$, because that's what the experiment usually finds.

How can that be, considering that 4 carriers are needed by definition to remove one Si atom? The answer is that some of those 4 carriers are not flowing through the external circuit but are used up in the electrolyte triggering a chemical reaction (with "chemical currents") that changes the electrolyte composition. A possible gross reaction as given in [1] could be

$$\mathrm{Si} + 4\, \mathrm{HF_2^{-}} + \mathrm{h}^{+} \Rightarrow \mathrm{SiF_6^{2-}} + 2\, \mathrm{HF} + \mathrm{H_2} + \mathrm{e}^{-} \,. \tag{2.8}$$

The valence of this process is obviously $n = 2$: one hole flows out of the Si and one electron is injected. Where are the chemical currents in this case? This is only revealed if one breaks down the gross reaction of Equation 2.8 into the series of actual step-by-step reactions that are really occurring. Figure 2.3 gives one of many possible versions; it takes into account that the stable end product of the dissolution process is the complex $[2H^+ - SiF_6^{2-}]$. The arrows symbolize the current flowing; solid and dashed lines denote positive or negative charges, respectively, the arrow thickness characterizes the amount. What happens, in short, is that a first hole breaks one Si bond of a Si surface atom; an electron that is injected breaks a second one (reaction 1). Both carriers contribute to the external current. The following reactions 2–5 bring the Si ion in solution as complexed $[H_2SiF_6]$.

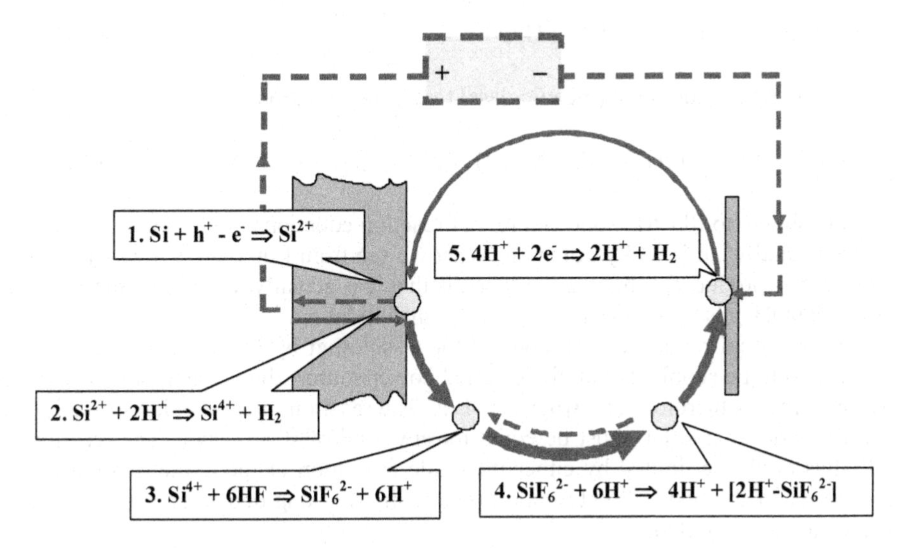

Figure 2.3. Detailed reactions with carrier and current flow for direct dissolution with a valence of $n = 2$

In total 4 positive charges "flow" from the Si working electrode to the counter electrode and 2 flow back, for a grand total of 2 charges flowing in the external circuit in order to dissolve one Si atom. In order to get some idea about the over-all reaction rate, it would be necessary to know, which one of the many detailed reactions listed (or the not listed ones) would be rate determining.

Adding up the reactions 1–5 gives the gross reaction equation for this case as

$$Si + 6\,HF + 1\,h^+ \Rightarrow [2H^+ - SiF_6^{2-}] + H_2 + 1\,e^- . \tag{2.9}$$

If this looks complicated, the reason is that it actually is complicated. The essential points to take note of are: i) Direct dissolution can have any valence between 1 and 4. In many cases, however, it will proceed with a valence around n = 2. ii) Direct dissolution consumes HF and liberates hydrogen (H_2). iii) The occurrence of H^+ in the detailed reactions in Figure 2.3 gives a strong hint that the reaction rate might be sensitive to the pH value of the electrolyte and that local reactions (e.g. at a pore tip) may locally change the pH value of the electrolyte.

2.3.3 Dissolution by Anodic Oxidation

Dissolution by anodic oxidation of Si followed by purely chemical oxide dissolution is a somewhat simpler process with respect to the valence, described by the gross reaction equations

$$Si + 2\, H_2O + 4\, h^+ \Rightarrow SiO_2 + 4\, H^+ \,. \tag{2.10a}$$

$$SiO_2 + 6\, HF \Rightarrow SiF_6^{2-} + 2\, H^+ + 2\, H_2O \,. \tag{2.10b}$$

Adding up produces the (meaningless) total gross reaction

$$Si + 6\, HF + 4\, h^+ \Rightarrow [2H^+ - SiF_6^{2-}] + 4\, H^+ \,, \tag{2.10c}$$

only shown to illustrate once more that reaction equations might not help much to understand what is really going on. What we see from Equation 2.10c is that no hydrogen is produced, what we don't see is that it is actually the water (more precisely the OH^-) that provides the oxygen for the oxidation.

The essential points to take note of for dissolution by oxidation are: i) Four holes need to be supplied from the external power source, the valence of this "oxidation" process therefore is **always** $n = 4$. ii) The reaction rates of oxide formation and of oxide dissolution must be equal for any kind of steady state. The reaction rate thus might be limited by either the oxide formation or the oxide dissolution; the slower of the two will determine the rate. iii) A strong dependence on the pH value must be expected.

As an additional point, we quickly consider the case that the oxide dissolution is the rate-determining step. If, for the sake of the argument, we let it be zero, we will experience pure anodic oxidation – only SiO_2 is formed. This, however, cannot possibly occur at some steady state because the oxide formed is an insulator and will impede current flow, i.e. the reason for its existence. In a potentiostatic experiment we would expect that the current will quickly decrease to zero; in a galvanostatic experiment we would expect that the external voltage goes up forever (or until a fuse blows) in order to drive a constant current through an increasing resistance. While this is roughly what will happen, anodic oxidation shows far more complex behavior if investigated in detail; cf. [22,23]. Going back to the case that SiO_2 is actually dissolved but at a rate far slower than its possible production rate, we must expect that the surface then is always completely covered by a thin oxide layer with some kind of (average) steady-state thickness. The applied potential then would drop to an appreciable amount in this oxide layer. Note that the steady state postulate above does not mean that production and dissolution rate of the oxide must be identical. For example, oxide can be produce very quickly by a large current flow in some area (Δx, Δy) on the Si surface (i.e. some "pixel" on the surface); after that it dissolves slowly with the current in that pixel temporarily shut off. An average over many uncorrelated pixels would give a steady state with a constant current. This is emphasized here because it is actually what happens during macropore etching, as we will see later.

2.3.4 Pore Formation

The big question is now: Pore formation requires localized dissolution – how does that tie in with the dissolution mechanisms discussed so far? What reason, exactly, has a perfectly flat Si electrode with some anodic current passing through it, to start **current oscillations in space**?

To appreciate this somewhat unusual way to look at pores in semiconductors (including Si), we look at Figure 2.4.

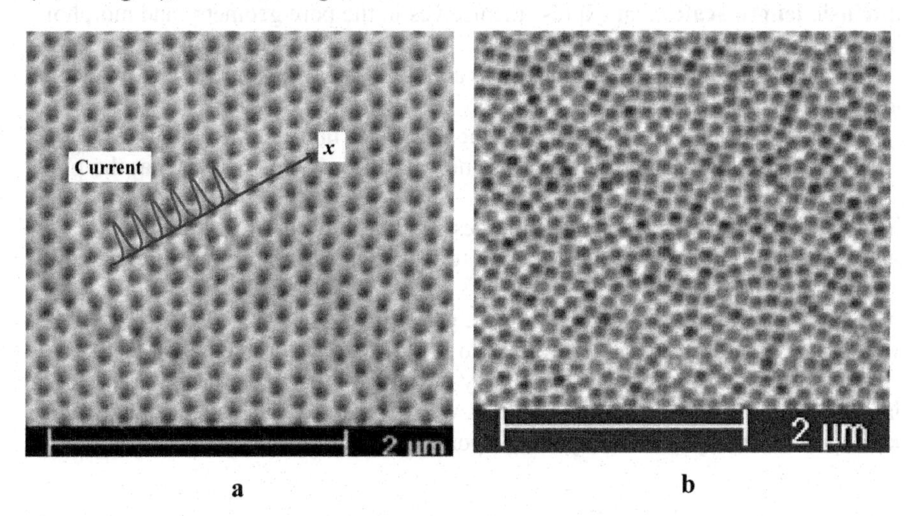

Figure 2.4. Self-organized pore crystal in InP, **a** and a self-organized frustrated pore crystal in Si, **b** illustrating the concept of current oscillations in space

What we see are deep macropores according to the IUPAC definition (see below) with diameters in the 100 nm region, arranged in a self-organized pattern. Pore etching in these cases started from a polished surface with no defined pore nuclei and involved formation processes relying on electrical break-down of the SCR at the pore tip. In the InP case shown in Figure 2.4a, self-organization actually produces a single pore crystal and the concept of current oscillations in space is obvious [24]. Pores grow only where current flows, in between pores the current is zero or very small. The current pattern thus is rather periodic in two spatial dimensions. The Si case in Figure 2.4b shows n-macro(org, breakdown, random) pores, i.e. pores obtained with (lightly-doped) n-type Si in the dark under junction breakdown conditions without lithographically defined nuclei (i.e. "random" nucleation). It is still possible to describe this pattern as current oscillations in space – just with a more complex but well-defined spectrum that is far from being random. The term "frustrated crystal" in this case refers to a peculiar property of these pores: the position of the nearest neighbors of a pore with respect to the orientation is random or uncorrelated, while the second-nearest neighbors are clearly correlated [25,26]. The structure, in other words, tries to be hexagonal and cubic at the same time.

Considering that in what follows we deal mostly with n/p-macro(litho) pores, i.e. with pore arrays defined by lithography, the issue of self-organization may seem to be of little importance. However, if the pattern expressed by lithography is too different from the pattern that would result from self-organization, the pores might "choose" not to follow the extrinsic nucleation pattern offered but to go for their intrinsic ordering scheme. Fortunately, Si is rather tolerant in this respect; this is, however, not the case for most other semiconductors. Nevertheless, the mere fact that pores result from some etching experiment necessitates the existence of **intrinsic length scales** that express themselves in the pore geometry and morphology.

If we enlarge upon the questions from above, we now have to deal with the topics:

i) Why do pores nucleate and grow in the first place under some conditions? How does Si dissolve at the pore tips but not (or at least much slower) at the pore walls?
ii) What kind of geometries, morphologies and aspect ratios are possible?
iii) How good is the achievable precision (e.g. in diameters or pore wall roughness) and the reproducibility?
iv) How can one control and monitor pore growth?
v) What are the limits, e.g. with respect to the growth rate or maximum depth?

These questions describe a large body in a huge parameter space that has still large unexplored regions. In what follows we will therefore focus on some particular important examples and only touch upon more remote issues.

2.4 A First Approach to Pores in Semiconductors

2.4.1 Basics

It is hard to describe the structure of porous semiconductors in words because we encounter a huge range of geometries and morphologies. The only (not very helpful) formal definition considers only the pore **geometry** in the sense of pore size (the word pore diameter is already to special because it implies a circular cross-section but we will use it anyway). According to an IUPAC convention [27] we distinguish between **micropores** (diameter < 2 nm), **mesopores** (diameter 2–50 nm), or **macropores** (diameter > 50 nm). However, the formally undefined term "nanopores" is also used a lot (for micropores), and so are names that relate to the morphology ("branched pores", tree-like pores, "nano-sponge") or the formation mechanism ("break-down pores"). More important, perhaps, is the fundamental distinction between "crystallographic pores" and "current-line" pores [4]. The former grow in some defined crystallographic direction (in Si, for example, either in <100> or <113>), the latter grow in the direction of the local current flow and therefore perpendicular to the equipotential planes in the semiconductor. Most meso- and macropores in all semiconductors fall into one of these two categories; which does not exclude, however the possibility of a third group: random pores, growing along some "random walk" direction. Mixtures are possible too, but we will not pursue this point here because for the case of Si we will encounter mostly crystallographic pores (in contrast to pores in some III–V or II–VI compounds). Note, however, that the fact that most pores in the standard {100} Si wafers grow in the <100> direction perpendicular to the surface does not prove their crystallographic nature – the current also flows perpendicular to the surface in this case.

There is no convention on the use of the term "**morphology**". In the context of this book we are mostly interested in the "straight-cylindrical" morphology, possibly with diameter modulations, but practically all other conceivable pore morphologies exist, too. As a kind of lumped parameter the **porosity** is often given, simply expressing the percentage of Si dissolved relative to the bulk in which pores were formed. Another indirect measure of morphology is the total **surface area** per cm^{-3}, where values of up to 1.000 m^2/cm^3 [1,2] have been found. Of course, higher porosities imply large surface areas, but the precise relationship between the two parameters depends on the detailed geometry and morphology (a few big pores may lead to the same porosity as a "nanosponge"; the latter, however, will have a far larger surface area).

For straight pores we are also interested in their aspect ratio (the relation of the pore length l_{pore} to its diameter d_{pore}) and perhaps in Δd_{pore}, the intended or unintended diameter variations.

For many applications of pores a precisely defined array on the Si surface (x, y plane) is needed. With standard lithography it is always possible to structure some masking layer in the desired way for geometries ranging from deep sub-µm structures to any µm structure and thus provide preferred nucleation sites for subsequent pore growth. It is important to note that in this way we only define the distance a between pores (or lattice constant for the typical periodic arrays) and therefore also the total number of pores in that part of the sample exposed to the etching. We do not define the pore diameter by lithography; this is rather an **internal degree of freedom** of the system. Consider an experiment where N pores are to grow and the global current density is set to j_G via a galvanostat. Assuming uniform etching (not at all a matter of course) each pore then grows with a local current density of $j_{Pore} = j_G / N$. According to Equation 2.1, the volume dissolved per time unit is $\Delta z \cdot \pi r^2$, and this is achievable by growing a thin pore rather fast (large Δz = gain in depth) or a fat pore slowly. It may even happen that the pores grow with oscillating diameters despite a constant external current, as we will see later.

What this means quite generally is: i) We need some clear way to control the pore diameter (or other properties) for a given a by adjusting some externally available parameters, and ii) we must expect that properties are related: Changing one parameter might influence many properties.

2.4.2 Parameter Space for Pore Etching Experiments

What kind of parameters do we have at our disposal? Quite a lot at the outset! The following classification gives a first overview for Si (with easily envisioned generalizations for other semiconductors); augmented by some general remarks. We start with the condition of the Si sample in Table 2.1.

From a practical point of view, there are only two sensible adjustable parameters. The first is the doping concentration, usually expressed in terms of the specific resistivity ρ. Ranges of ρ = (100.000–0.001) Ωcm are available; for the standard n-macro(aqu, bsi, litho) pores values of ρ = (100–0.1) Ωcm are most commonly used. The second is the illumination intensity P. Typically, high P values are needed in order to inject sufficient currents. In order to have a rough feeling for the quantitative correlation between illumination intensity and injected photocurrent j_P, we may note that the most intense sunlight experienced on this planet produces $j_P \approx 35$ mA/cm^2. Since only a fraction of the photo generated holes will reach the pore tip and convert to etching current, suitable light sources (especially for large areas) for photo currents in I this order of magnitude require some involved engineering; we will come back to that.

All other parameters listed above simply should be optimal; they cannot be used to adjust pore properties. A large diffuse length is a necessity if enough holes are to reach the pore tip; even more important is that the diffusion length is uniform in (x, y).

The backside surface merits special consideration. In any case – for p-macro(org.) (i.e. macropores grown in an organic electrolyte in the dark) or for the standard n-macro(aqu bsi) pores, the electrical contact has to be good and uniform; especially for large samples. This is easily achieved by a good standard metal contact (involving p^+ or n^+ implantation, defect annealing, cleaning, Al sputtering, Al sintering, and passivation), the necessary processes, however, will almost certainly "kill" the diffusion length or make it non-uniform if done without special precautions. If backside illumination is to be used, nothing can be deposited on the backside that absorbs the light, excluding any metallization. The only remaining possibility is the formation of an n^+ layer, which must be done without compromising the diffusion length and without making the back side recombination active (i.e. keeping the back side recombination velocity S_B small and uniform is important). The reasons are clear: Backside illumination generates the important holes close to the back side surface. Any hole lost by recombination at the backside surface or in the bulk cannot contribute to the dissolution of Si at the pore tip. As long as the losses by recombination are uniform, they can be compensated to some extend by increasing the illumination intensity P. If there is some non-uniformity either in L or S_B, current flow through the pores will be non-uniform too, leading to unavoidable variations of pore parameters over the area (x, y).

Table 2.1. Parameters of the Si sample to be etched that determine or influence the pore properties

Silicon "input" parameters	
Parameter	**Remarks**
Doping type	n- or p-type; makes all the difference!
Doping concentration N_D. (or resistivity ρ in Ωcm)	Doping defines the width d_{SCR} of the SCR, one of the most important **intrinsic length scales**. Etching behavior and pore properties change completely at high doping levels ($\rho <\approx 0.01$ Ωcm).
Minority carrier life time τ (or diffusion length $L = [D\ \tau]^{1/2}$)	Governs how many photo-generated holes will be able to reach the tip of a growing pore from wherever they have been generated.
Illumination status	Usually back side illumination (bsi); rarely front side illumination (fsi).
Crystal perfection	Uncritical in Si (therefore large τ).
Surface orientation	Usually {100}; occasionally {111} or others. Defines pore growth direction in most cases.
Surface conditions of front side	Usually polished and clean on front side; might have defined nucleation site by masking. Type of mask is critical for prolonged etching.
Surface conditions of back side	Might be critical for bsi; for details see below.
Size	The difference between etching a 1 cm^2 sample or a 300 mm wafer, beside know-how, amounts to about € 400.000.– in equipment.

Next we look at the **chemical parameters**; essentially the electrolyte composition.

One glance at Table 2.2 is enough to realize that the parameter space with respect to the chemistry is approaching infinity. In what follows we use the simple distinction given in [3] and classify electrolytes into "aqueous", "organic" and "exotic". The major rules of thumb in this context are:

i) The HF concentration always limits the flow of current and thus determines the maximum etching or speed or pore growth rate.
ii) The "oxidizing power"; loosely the ability of the electrolyte to produce SiO_2, has major bearing on pore formation and properties; cf. [28] for more details on this point. Obviously, electrolytes with little water or simply no oxygen bearing species have small oxidizing powers.
iii) Organic electrolytes thus generally decrease the oxidizing power (less water and therefore OH^- available for the oxidation, cf. Equation 2.10a. This is an essential condition for the formation of macropores in p-type Si.

Table 2.2. Parameters of the electrolyte that determine or influence the pore properties

Chemical parameters (electrolyte composition)	
Parameter	**Remarks**
HF; F^-, HF_2^-, ..	Absolutely necessary. Typically from aqueous HF at 49 wt. %. Occasionally NH_4F dissolved in water.
Major solvent	H_2O, Ethanol (EtOH), other alcohols, organic solvents like DMSO, DMF; Acetronitrile, …, mixtures.
Minor additions	Surfactants; anodic, cathodic or neutral.
Major additions	Acetic acid, conducting salts, viscosity enhancers, …
Oxidizer additions	H_3PO_4, H_2O_2, CrO_3, ... Typically for macropore etching in n^+–Si.
Composition	Often HF : H_2O : EtOH with [HF] = 3 % … 30 % and at least 10 % EtOH. Otherwise (almost) any combination of all of the above at (almost) any concentration ratios has been used.

iv) Adding oxidizers does the opposite; they are needed, e.g. for producing macropores in n^+-type Si [29].

v) Surfactants, viscosity enhancers, addition of "conducting" salts to increase the conductivity of the electrolytes and so are supposed to make pores more "perfect" (e.g. with respect to uniformity, or the pore wall roughness) but should not change the general etching behavior [30,31].

There is a wealth of data on the various electrolytes contained in the possible set defined in Table 2.2 with respect to pore etching, but not much predictive power beyond the rules of thumb given above. In other words: Optimizing the etching of some specified pores needs a more or less empirical optimization of the electrolyte, and this can be an unending task. In what follows a recent example for this is given and discussed in some detail, so we will not belabor this point here any more.

Finally let's look at the **system parameters** (see Table 2.3).

Table 2.3. System parameters that determine or influence the pore properties

System parameters	
Parameter	**Remarks**
Darkness: Voltage and current	One degree of freedom: Potentiostatic conditions: $V(t)$ is given; I adjusts. Galvanostatic conditions: $I(t)$ is given, V adjusts.
Illumination (bsi / fsi): Voltage current and illumination intensity	Two (limited) degrees of freedom allow several modes. Typical for bsi: $I(t)$ and $V(t)$ are given; $P(t)$ adjusts via feed-back loop.
Geometry and type of counter- and reference electrode	Important; in particular for large area etching.
Electrolyte flow conditions	Crucial for uniformity and to some extent for pore parameters.
Temperature	Large influence on pore parameters; should be controlled to at least +/– 0.5 °C. Generally better results at low temperatures (down to –40 °C); optimization may need programmed $T(t)$.
Illumination	Large and uniform intensity is required without heating the system too much. Critical for large area etching; requires LED matrix with ≈ 2.000 LEDs for 150 mm wafer and precise control of P(t).
Back side contact	Either mechanical or by electrolytic double cell; each with its own set of problems. Especially crucial and demanding for large area etching.
Handling of gas evolution	Hydrogen bubbles must be avoided at all costs; the gases produced must be safely removed.
In-situ monitoring and feed-back control loop	Just emerging; see below.

Like before, we have only voltage V, current I, and illumination intensity P as truly adjustable parameters (the temperature does not really count in this context) and in a real pore etching experiment we can at best fiddle with one parameter in the dark and with two for illumination. The rest must simply be optimized.

Many of the parameters enumerated fall under the heading "cell design". This is a major issue, in particular for large samples, and it must be mentioned here that all electrolytes are extremely dangerous and corrosive. The HF always present in the electrolyte excludes the use of glass, and organic electrolytes will destroy almost any materials within weeks, demanding 100 % Teflon constructions.

We have now established a certain framework for pore etching; next we will look at the standard n-macro(aqu, bsi) pore case, keeping the questions posed at the end of Section 2.3.4 in mind.

2.5 The Etching of Deep Macropores in Lightly Doped n-Type Si Under Backside Illumination

Figure 2.5 shows schematically the basic scenario for n-macro(aqu, bis) pore etching at a point in time where the pores have reached a depth or length l_{Pore}. Within the so-called space-charge region model proposed by Lehmann and Föll in the first publication concerning macropores in Si, [32], it is easy to understand why these pores form at all. Figure 2.5 shows the model in a slightly modified version of the original graphics. In essence two points conspire to produce macropores: i) Whatever the detailed dissolution reaction according to the equations in Section 2.3.2 will be, the most important rate limiting factor is the concentration of holes available for the process; in all variants at least on hole is needed to trigger the process (cf. the details in Figure 2.3). For n-type Si (or any other semiconductor for that matter) it means that if holes are only supplied at pore tips (or at pre-defined nucleation points), pores will form and grow.

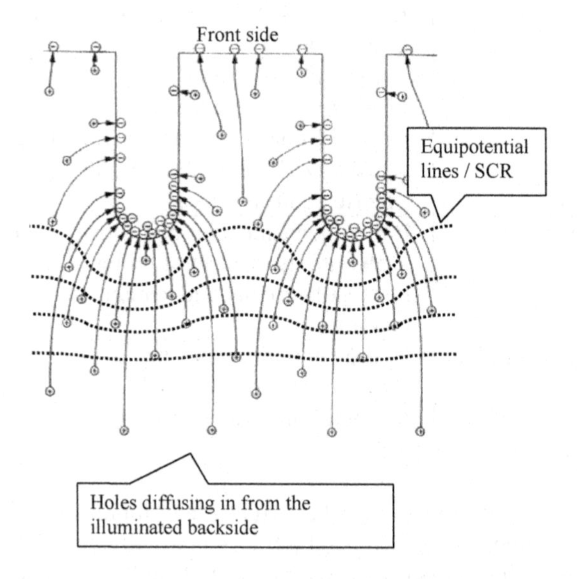

Figure 2.5. SCR focusing model of n-macro(aqu, bsi) pore growth

There are two basic ways to achieve this: i) Use the stronger electrical field within the space charge region that is present at concave surface curvatures (like the tip of a pore) to generate holes via some electrical breakdown mechanism. ii) Use the curvature of the SCR around pores or concave depressions to focus holes coming from the backside on the pore tip. In passing we may note that the first possibility is conceptually and technically easier than the second (just crank up the voltage until substantial current flows, i.e. electrical breakdown occurs, and you will find pores). However, most likely mesopores will be obtained, hard to control (breakdown is a very non-linear process) and with geometries and morphologies not suitable for most optical applications. It is thus the second possibility that is used for good n-macro(aqu, bsi) pores. The current flowing through the pore tips is just a fraction of the total photo current j_P; only if the diffusion length L is at least in the same order of magnitude as the distance between the pore tips and the back side, an appreciable fraction of the photo-generated holes will reach the pore tip. This limits the "bsi" method to Si because the necessary diffusion lengths in the order of > 100 μm are not possible in all other semiconductors with the exception of possibly Ge, where, however, it did not work so far [33].

The SCR focusing model thus explains why pores form and grow; it does not, however, make direct statements about the pore geometry, i.e. pore diameter d_{pore} and distance/lattice constant a. Indirectly it is clear from Figure 2.5 that a should be in the order of twice the space charge region width d_{SCR} in order to keep the porous part of the substrate field free. Moreover, the pore diameter must be larger than some critical diameter d_{BD} at which the curvature of the pore tips is so large that the field strength on the SCR is sufficient to trigger avalanche breakdown. It is also clear that these points relate directly to the doping concentration N_D via

$$d_{SCR} = (2\varepsilon\varepsilon_0 V / e\, N_D)^{1/2} \tag{2.11}$$

with $\varepsilon = 11.9$ = dielectric constant of Si, and ε_0 = permittivity of vacuum.

There is no simple equation correlating the breakdown voltage V_{BD} to the pore radius r or more precisely to the radius of curvature of the pore tip, with the resistivity as parameter, because there are at least two different breakdown mechanisms (tunneling and avalanche effect; cf. [21]) and the exact shape of the pore tip might be of importance. From the graphs given in [1] one can deduce as a rule of thumb for the critical pore radius r_{BD} where break down occurs

$$V_{BD} \propto r_{BD}\,, \tag{2.12}$$

for $r_{BD} < 100$ nm. To give examples: For pore tip radii of $r \approx 10$–20 nm breakdown occurs for voltages around 1 V for the complete typical range of doping ($N_D \approx (10^{15}$–$10^{18})$ cm^{-3}); for heavily doped Si ($N_D \approx 10^{18}$ cm^{-3}) breakdown always occurs at voltages below a few Volts even for large r_{BD}, whereas for lightly doped Si more than 100 V would be needed for $r \approx 1$ μm.

If one now imagines a macropore etching experiment for a sample with area A at a voltage as low as possible but in the saturation part of the IV characteristics as shown in Figure 2.2 with a global current density j_g, and a lithographically defined periodic structure of nucleation points (typically with a cubic or hexagonal symmetry) with a lattice constant a, the number of attempted pores is fixed at $A / A_{lattice}$ with $A_{lattice}$ = area of elementary cell of the nucleation point lattice (e.g. $A_{lattice} = a^2$ for a square lattice). The current density relative to A is j_g, but the current density for a circular pore with diameter $d_{pore} < a$ by definition, and the current density etching a cross-sectional area of $A_{pore} = (\pi / 4)\, d^2_{pore} < a^2$ must be concomitantly larger. The pore, in other words, has a degree of freedom to choose a current density j_{pore} that can be larger than j_g; it is only limited by the minimum diameter dictated by breakdown and by how much current the tip can actually process. It is reasonable to assume that the pore "goes for" the only current density value that has some special significance, and this is the peak value j_{PSL} featuring prominently in the initial *IV* characteristics. This is akin to stating that the pore maximizes the current density for the conditions given. The late Volker Lehmann made this assumption in 1993 [34], which leads immediately to the "Lehmann equation"

$$j_{PSL} \cdot (\pi / 4)\, d^2_{pore} = j_g \cdot A_{pore}\,, \tag{2.13a}$$

$$d_{pore} = [(4\, A_{pore} / \pi) \cdot (j_g / j_{PSL})]^{1/2}\,. \tag{2.13b}$$

It is remarkable how well the SCR focusing model together with the Lehmann formula (now called SCR model for short) works, considering that there are many weak points that will be raised later. Here we note some immediate connotations:

i) The concept works even for random pores, i.e. without defined nucleation points provided lithographically. The self-adjusted distance between the pores will be in the order of twice d_{SCR} since it can't be much smaller; if it would be larger new pores will nucleate between too distant pores, see [35,36] and thus some average $A_{lattice}$ exists; it is, so to speak, the drainage area for one pore from which to collect the current. It is of interest in this context that random macropore nucleation is a tortuous process that takes roughly an hour and goes through several stages before reaching some kind of steady state, cf. [36].
ii) Pores with a large drainage area $A_{lattice}$ – by happenstance for random nucleation, or for pores with no neighbors in lithographically designed patterns – will have large diameters. Pores at the edge of some array of nuclei with limited size will have far larger nominal $A_{lattice}$ values than regular ones and will tend to branch into the direction of the hole supply. This is a severe limitation on the kind of possible pore arrays; cf. [37].
iii) If the global current density is modulated, e.g. sinusoidally, the pore diameter should show the same kind of modulation (within limits, of course). This is essential for many applications including the omnidirectional filters discussed in Chapter 11 but also a major point that will make the limits of the SCR model obvious.
iv) The growth speed $v_{Pore} = dz / dt$ of the pore is directly determined by j_{PSL}; cf. Equation 2.1b. To give an example: For $j_{PSL} = 10$ mA/cm^2 (a typical value for the typical HF concentration of [HF] = 5 wt. %) a pore diameter of 1 μm and a "drainage area" of 10 μm^2 (cubic lattice constant of ≈ 3 μm), we have $v_{Pore} = 1$ μm/min; for [HF] = 30 wt. % with a $j_{PSL} \approx 150$ mA/cm^2 one could expect $v_{Pore} =$ 12 μm/min.
v) For not immediately apparent reasons, the numbers given above for [HF] = 30 wt. % are somewhat virtual. A general experimental finding has been that no deep pores can be etched for HF concentrations above roughly 5 wt. %; this will be discussed in detail in what follows.

vi) The SCR model does not contain direct information about the maximum depth of the n-macro(aqu, bsi) pores achievable. There is, however, indirect information: the potential at the pore tip will decrease with increasing depth because of ohmic losses, and the concentration of the chemical species will decrease for educts, and increase for products due to diffusion limitations. In essence, this lowers the j_{PSL} value as a function of depth and this can be compensated for to some extent by lowering j_g in order to keep the j_g / j_{PSL} ratio and thus the pore diameter constant. Lehmann has provided some equations for that [34] that works remarkably well, if properly implemented even in parts of the parameter space where its application is doubtful.
In principle, raising the applied potential and the HF concentration with time would also be an option to compensate these "depth losses" – but that doesn't work! We will belabor this point in what follows in some more detail.

vii) The very first illustration given to the SCR model (see Figure 5.1a) already indicates a major problem not explicitly contained in the model: some holes, either coming from the back side or (thermally) generated nearby, will also flow through the pore walls. In other words: there are pore wall currents I_{wall} in addition to the pore tip current about which little can be said from general principles at this point except that they can be expected to increase linearly with the pore wall area, i.e. linearly with time. These pore wall currents must be seen as major problem as will become clear from what follows.

A large number of n-macro(aqu, bsi) pores has been produced since their discovery in 1990 [32] and the reader can find many pictures and details in numerous publications (see e.g. references in [3]) and in the remainder of this book, so we will not dwell any more on what we will call now n-macro(aqu, bsi, classical) pores. Instead some room will be given to recent developments concerning in-situ monitoring and what can be learned from first results.

2.6 In-Situ Monitoring of Pore Growth

2.6.1 Motivation and Experimental Approach

From the preceding sections it became apparent that anodic pore etching in semiconductors is a rather complex issue because it happens at the intersections of (electro)chemistry, semiconductor physics and some non-trivial process technology. Just hinted at in Figure 2.2 was the fact that there are also pronounced self-organization features, meaning that one needs to add the physics of critical phenomena to the list. The fact that pore etching in semiconductors, despite major progress in the last 15 years, is still a bit of a "black art" as soon as one deviates from well-explored areas in parameter space, has one of its roots in this multidisciplinary confluence of topics that are difficult enough in their own right.

An independent problem plaguing pore etching is the lack of in-situ data. If we look, for example, at a typical n-macro(aqu, bsi, classical) pore etching experiment that lasts for up to 10 hours, all the information we have in-situ is how the one free parameter (here typically the light intensity $P(t)$) adjusts to meet the pre-defined values of I and V. Otherwise there is only ex-situ information obtained by examining the pore structure obtained, weighing the sample to determine the total amount of Si dissolved and thus the global valence $n_G = <n(t)>$, checking the wall roughness by, e.g. atomic force microscopy, and so on. Of course, measuring the desired device performance, e.g. optical filter characteristics, may also be a way to assess the ex-situ or "output" parameters of the etched sample.

It is clear that it would be desirable to have some direct in-situ information about what is going on at the growing pores. This would not only be helpful to control the etching process (ideally automatically, via a feed-back loop and a model) but could also provide data that can be tested versus hypotheses concerning the etching process itself. There is, however, no obvious monitoring technique out of the long list of possible methods (cf. e.g. the overview of Chazalviel [38]) because it is not so simple to "look" down a very long pore at the interface Si/electrolyte where things happen. The only method that can make some claim to in-situ monitoring is dual-mode FFT impedance spectroscopy (IS) as recently published in [39].

Impedance spectroscopy is a "black box" method that monitors the response of a system output to small disturbances in the form of a $A' \cdot \sin(\omega t)$ modulation at its input with A' = (small) amplitude. Linearity is assumed, i.e. the output can always be written as $A'' \cdot \sin(\omega t + \varphi)$; i.e. the measured quantities are output amplitude A'' and phase shift φ as a function of the circle frequency ω. For small enough amplitudes the system can always be linearized; small amplitudes are also necessary in order to keep the pore properties unaffected.

Applying the technique to n-macro(aqu, bsi) pores is easy in principle but not in practice. First of all, the response comes from whatever happens inside of typically millions of pores, and if etching is not very uniform no strong conclusions can be drawn from data that then represent some unclear averages over millions of pore tip (and pore wall) properties. This problem can be overcome with carefully optimized etching conditions.

Second, measuring many frequencies subsequently may simply take too long – the pore parameters, e.g. its depth, may have changed too much during the measurements. Modulating the input with all the desired frequencies simultaneously plus performing a (fast) Fourier Transform of the output (FFT) solves this problem. However, because the total amplitude of a signal containing many frequencies now is larger, linear response demands smaller individual amplitudes leading to signal/noise problems. Nevertheless, FFT impedance spectroscopy is a must for fast in-situ pore growth monitoring.

Last, classical impedance spectroscopy, typically modulating the voltage and monitoring the current (abbreviated *IV*–IS in what follows) might provide good amplitude/phase data but it is not clear offhand what those data mean. Without a quantitative model for the impedance $Z_{IV} = dV/dI$ that contains the primary parameters of interest against which the measured data can be fitted, parameter extraction is either not possible, restricted to indirect parameters, or speculative – if not actually questionable. A model for *IV*–IS essentially would have to give the *IV* characteristics and, as pointed out before, such a model does not yet exist. This problem can be overcome to some extent by using "equivalent circuit" models, but major progress comes from resorting to a new mode of IS: modulate the intensity $P(t)$ of the back side illumination and record the response in the current. As it turns out, the pseudo-impedance $Z_{bsi} = dP/dI$ (the term "impedance" is reserved for quantities measured in Ohm) can be derived theoretically in great detail. This mode will be called bsi-IS. Using both modes finally establishes the dual-mode FFT IS technique mentioned in the beginning.

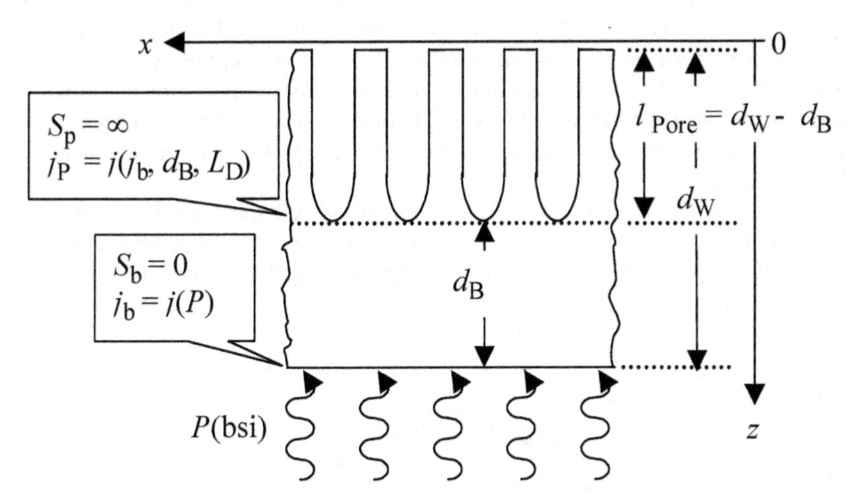

Figure 2.6. Geometry and definitions needed for bsi-IS. See text for details

Dual-mode FFT impedance spectroscopy is a demanding technique and we will not go into great details discussing its finer points here (cf. [40,41] for details). Assuming that meaningful raw data can be obtained, Equations 2.14–2.16 together with the definitions contained in Figure 2.6 provide a secure base for parameter extraction:

A complete theoretical impedance for bsi-IS relative to n-macro(aqu, bsi) pore etching is given in Equation 2.14

$$Z_{\text{complete}}(\omega, d_B, L, \Delta S_P) = A_1 \left(Z_{\text{bsi-1}} + \frac{A_0}{\frac{1}{Z_{\text{bsi-1}}} + A_2\sqrt{i\omega}} \right) \left(\frac{\frac{1}{\sqrt{\frac{1}{L^2} + \frac{i\omega}{D}}}}{\frac{1}{\sqrt{\frac{1}{L^2} + \frac{i\omega}{D}}} + \Delta\frac{D}{S_p}} \right) \quad (2.14a)$$

$$Z_{\text{bsi}-1}(\omega, d_B, L) = \frac{dj_{\text{sem}}(\omega, d_B, L)}{dP(\omega)} \propto \frac{1}{\cosh\left(l_{\text{Pore}}\sqrt{\frac{1}{L^2} + \frac{i\omega}{D}} \right)}. \quad (2.14b)$$

Three contributions can be distinguished: The first and the second large bracket in Equation 2.14a and the quantity Z_{bsi-1} defined in Equation 2.14b. The latter is the complete solution of the diffusion problem where a given number of holes is generated at the specimen's back side by illumination, diffuse through the sample with a diffusion coefficient D and recombine either in the bulk (described by the diffusion length L) or are converted to current a the pore tips. The later process is mathematically described by assigning an interface recombination velocity $S_P = \infty$ (or rather $1 / S_P = 0$) to a fictitious interface defined by the pore tips as shown in Figure 2.6. Within the SCR model this should be a reasonable approach because all holes reaching the position of this fictitious interface would end up at a pore tip and thus disappear.

However, as shown in Figure 2.7, where a measured spectrum is printed in a conventional Nyquist plot (the (negative) imaginary part of the impedance is plotted vs. the real part), matching bsi-IS data against just Equation 2.14b does not produce a perfect fit.

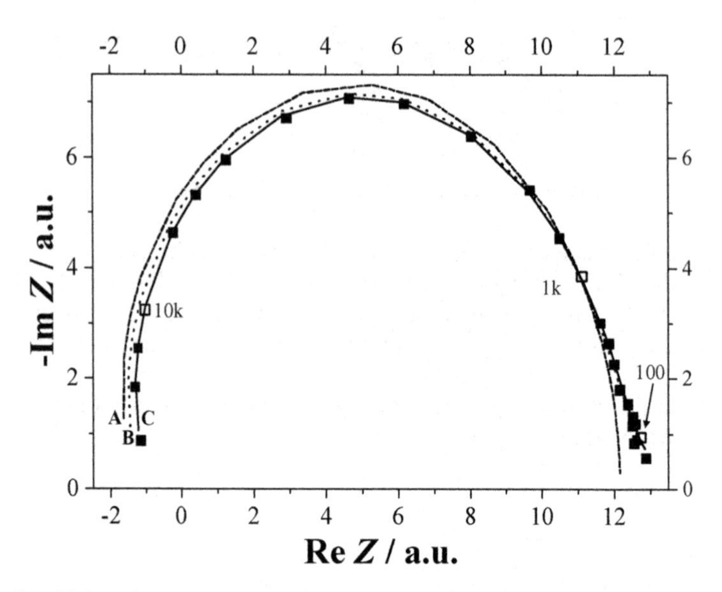

Figure 2.7. Fitting theory to measured in-situ bsi-IS data (black squares). Curve A results from Equation 2.14b only. Curve B results from Equation 2.14a without the first term, and curve C is based on the complete impedance as given in Equation 2.14a. Numbers indicate some of the 27 frequencies used in Hz (total range 10–40 kHz)

For sake of simplicity and clarity, Equation 2.14 does not take into account the backside recombination velocity (i.e. $S_B = 0$ in Figure 2.14) and the absorption constant α of the LASER light, since this is not very important. However, for the fitting of the impedance data shown here, these parameters are included and the full equations have been used with appropriate numbers for the two additional parameters. The problem of an imperfect match, however, still remains. In conclusion, the SCR-model for n-macro(bsi) pore growth is too simple and must be augmented.

At a next step of sophistication we now assume that some holes arriving at the depth l_{Pore} of the pores do not all contribute to the pore tip current but are also consumed at areas not very close to the pore tip; i.e. at pore walls. Mathematically this is expressed by changing the boundary condition $S_P = \infty$ in Figure 2.6 to a periodic expression where the recombination velocities between the pore tip and the point halfway between the pores differ by some ΔS_P. Taking this into account, the solution of this diffusion problem gives as a second approximation Z_{bsi-1} times the second bracket in Equation 2.14a that contains as a new parameter ΔS_P in the form $\Delta(D/S_P)$. The match to all the experimental data is better now (note that just on pore etching experiment yields several 100 bsi-IS spectra) but still not perfect.

A perfect match can be obtained in a third approximation if one assumes that some fraction of the holes reaching the pore tip region is not immediately consumed at the tip but has to "wait" a bit, meanwhile diffusing around. In physical terms this means that oxide covered parts of the pore tip cannot process holes instantly but need to wait until the oxide thickness has been sufficiently reduced. More succinctly stated (and with knowledge of quantitative data some of which are shown below), at this point it becomes necessary to postulate that the pore tip is actually covered with (thin) oxide most of the time.

The proper term for this "waiting time" presently cannot be obtained from solving an appropriately changed diffusion equation but is derived by adding a general diffusive term akin to what is known as "Warburg impedance" in *VI*-IS [42]. The quotation marks refer to the fact that "proper" impedances are measured in Ohm [Ω], as noted before. Besides the parameters L, l_{Pore}, ΔS_P (always found as D/S_P and therefore expressed as $\Delta(D/S_P)$), two new parameters A_0 and A_2 emerge beside A_1 that is just a proportionality constant and simply a measure for the over-all etching area defined by the envelope of the pore tips, including the area of under-etching at the O-ring or under masks. The combination of A_0 and A_2 quantifies the fraction of that part of the etching current that experiences the additional diffusion around the pore tips.

Evaluating the in-situ FFT bsi IS data thus should give in-situ information about the pore depth l_{Pore}, the diffusion length L, the total active area A_1 (essentially reflecting the degree of under etching), the detailed current flow processes around the pore tips quantified by A_0, A_2, and $\Delta D/S_P$, and the growth speed v_{tip} of the pore via $v_{tip} = dl_{Pore}/dt$. Since L is a constant by definition, the numerical evaluation procedure determines L from the first few spectra and then takes it as constant, making subsequent data extraction easier and more precise. Since $l_{Pore}(t)$ is now a known quantity, all the other parameters extracted can be rescaled from a function of time to a function of pore depth.

In the case of the *VI*-IS, the general approach is to find the best fit to equivalent circuits expressed either in suitable circuit diagrams containing capacitances, resistors and Warburg impedances or, slightly more abstract but fully equivalent, by using model equations describing processes with as many time constants as needed; cf. e.g. the standard text book of MacDonald for details [42]. The equation that matches many thousands of in-situ *VI*-IS obtained so far rather well is

$$Z_U(\omega) = R_s + \frac{1}{\left(\frac{i\omega\tau_{slow}}{(R_p+\Delta R_p)(1+i\omega\tau_{slow})} + \frac{1}{R_p(1+i\omega\tau_{slow})}\right) + i\omega C_p} . \qquad (2.15)$$

Following the standard interpretation of such a model equation, R_s describes the ohmic losses due to a general series resistance, C_p describes the capacitance of the interface, and R_p the so-called chemical transfer resistance of the chemical dissolution process. The chemical dissolution splits up into two processes with different reaction rates characterized by the relaxation time τ_{slow} of the slow process and the time constant R_pC_p of the fast process. The difference ΔR_p describes the increase in the chemical transfer resistance at higher frequencies. A similar approach has been used successfully for unraveling pore formation in InP, cf. [43,44] for details.

The processes in question can only be the current-driven direct dissolution of Si (typically occurring with a valence $n_{dd} = 2$) and current-driven Si oxidation (valence $n_{Ox} = 4$) together with the purely chemical dissolution of the oxide as detailed in Section 2.3.1. Assuming that direct Si dissolution is the "fast" process and that Si dissolution by oxide formation plus oxide dissolution is the "slow" process, it is possible to derive an equation for the over-all valence n (for a more detailed derivation see [39])

$$n = \frac{4}{2 - \frac{\Delta R_p}{R_p + \Delta R_p}} . \qquad (2.16)$$

A full evaluation of in-situ FFT *VI*-IS data thus provides as a function of time (or pore depth) the following parameters: R_s, R_p, C_p, τ_{slow}, ΔR_p, and as a combination of those parameters the valence n calculated from Equation 2.16.

Summing up, in-situ multimode FFT IS allows keeping track of 12 parameters that encode more or less directly properties of the Si (e.g. its diffusion length L), the actual pore geometry (e.g. the pore depth l_{Pore} or the degree of under-etching at O-rings or masked parts of the sample via A_1), the processes at the interface (e.g. the valence n or ΔS_P) plus some "electrical" properties like the series resistance and interface capacitance, and thus crucial information not only for controlling pore growth, but for any modeling effort that endeavors to get a better understanding of the process.

It is beyond the scope of this chapter to discuss in detail what kind of conclusion with respect to the pore etching process can be drawn from multi-mode in-situ FFT IS and we will restrict ourselves to two examples of particular interest for the next sections.

2.6.2 Examples for Multi-Mode In-Situ FFT Impedance Measurements

Figure 2.8 shows some n-macro(aqu, bsi, classical) pores, not very deep and not particularly impressive, together with plots of the time dependence of the ten most important parameters from in-situ multimode FFT IS. The FFT impedance spectroscopy measurements have been performed by a system produced by ET&TE GmbH. Measurements were taken at 1.5 s intervals, alternating between *IV*-IS and bsi-IS The signal contained 27 frequencies in the range between 50 Hz and 20 kHz.

Without going into details, the 10 parameters shown develop as one might have expected. After an initial nucleation phase has ended, most parameters change slowly and smoothly with the pore depth. In particular, the valence n is rather constant around 2.8 (its average $<n(t)>$, however, would be somewhat smaller) and increases slightly. This not only validates the assumptions made for the derivation of Equation 2.16 but demands that the pore tip is covered with oxide most of the time if the data are to be trusted. The growth speed v_{tip} of the tip is also rather constant around 0.8 μm/min, decreasing just a bit with time. The parameter $\Delta(D/S_P)$, after going through a maximum at a pore depth of about 15 μm, decreases smoothly thereafter. This is of some interest because it indicates a decreasing "focusing" efficiency of the pore tip with respect to the holes reaching it. The transfer resistance of the dissolution process increases smoothly, indicating diffusion-limited growth of the pore. The parameter A_1 increases continuously after nucleation, indicating that under-etching continues and does not (yet) saturate. Generally, all parameters are well behaved and change smoothly with depth as would be expected for pore growth to a relatively small depth under these conditions. Note, however, that there is no "steady state" in the sense that transport parameters stay constant.

Next we look at data from experiment that produces almost identical pores but with completely different parameter evolution with time as shown in Figure 2.9.

The major difference to the experiment shown in Figure 2.8 is the addition of 0.83 g/l carboxymethylcellulose sodium salt (CMC), added for a considerable (but unspecified) increase of the electrolyte viscosity. As noted in Section 2.4.3 increasing the viscosity of the electrolyte tends to make pores more perfect [30,31]; not only in Si but also in the rather different case of forming porous TiO_2 [45]. The reasons for this are unclear and we will not enter into speculations about possible mechanisms here.

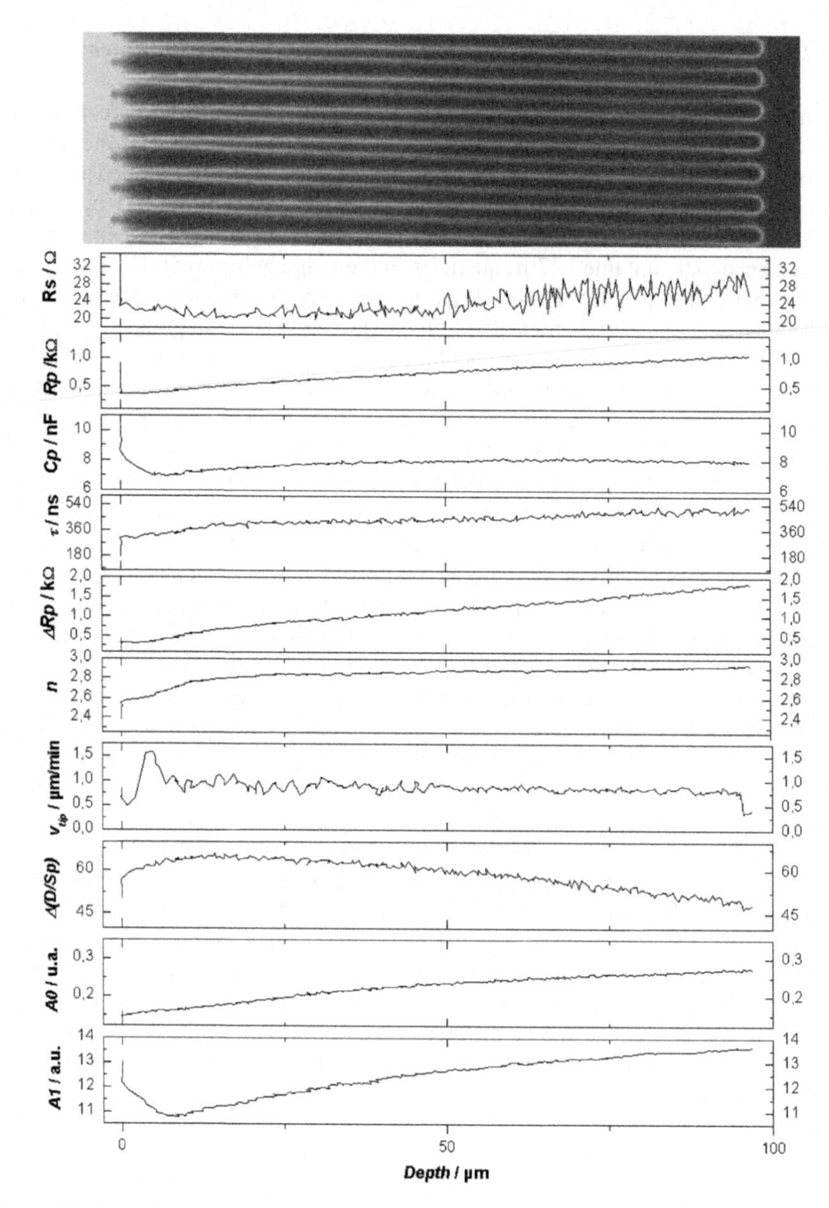

Figure 2.8. Impedance parameters vs. pore depth in direct comparison to the etched pore morphology for n-macro(aqu, bsi, classical) pores. A standard aqueous electrolyte with an HF concentration of 5 wt. % was used. The applied voltage was 0.4 V, and the etching current decreased from 21 mA/cm^2 to 18 mA/cm^2 during the total etching time of 126 min. Further data: T = 20 °C, n-type Si, (100), 5 Ωcm, pre-structured by standard photolithography, hexagonal lattice a = 4.2 µm

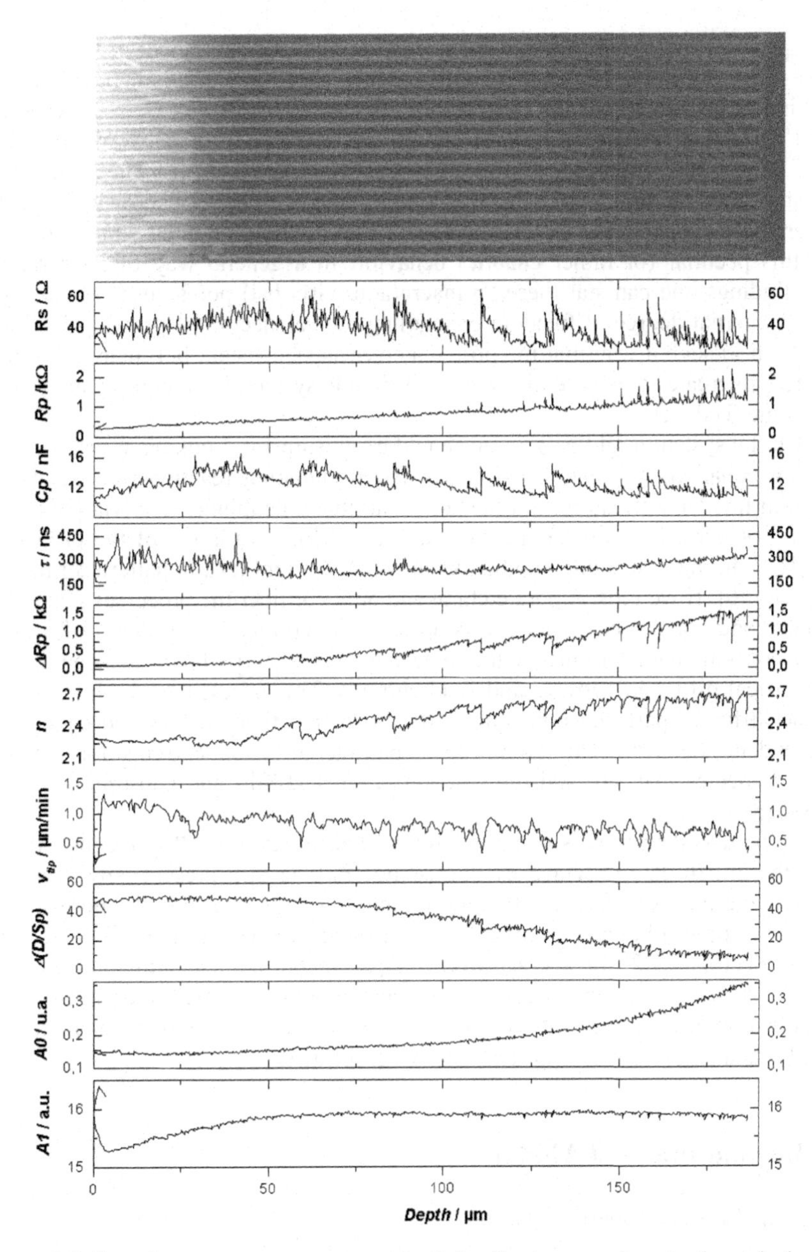

Figure 2.9. Impedance parameter vs. pore depth in direct comparison to the etched pore morphology. An aqueous electrolyte with an HF concentration of 5 wt.% and 0.83 g/l of CMC was used. The applied voltage was 0.7 V and the etching current decreased from 15 mA/cm^2 to 10 mA/cm^2 during the total etching time of 295 min. Further data: T = 20 °C, n-type Si, (100), 5 Ωcm, pre-structured by standard photolithography, hexagonal lattice a = 4.2 µm

Impedance measurements, however, provide hard data as shown in Figure 2.9. At least five out of the 10 parameters plotted now show very unruly behavior; there is a kind of periodic "pulsing", expressed, e.g., as sudden decreases in the valence n and correlated (negative) peaks in the growth rate v_{tip}. In other words, roughly every 20 μm the pores stop to grow or at least become sluggish for a while, then they start growing again. Note that this kind of behavior does not show in the post-mortem pore picture and that without in-situ data it would be impossible to recognize this peculiar (or rather chaotic) behavior. In a general way of interpreting these findings one can stat these "n-macro(aqu, visc, bsi) pores "use" the always present internal degree of freedom of juggling the valence n, the growth speed v_{tip} and so on in such a way that the pore diameter stays constant but many other parameters fluctuate. This is reminiscent of a chaotic system, i.e. a non-linear system with some feedback.

That the system might be chaotic should be no surprise. Looking at just the few reaction equations given in Section 2.3 and formulating reaction rate equations based on this would lead to non-linearity with many couplings (e.g. via $[H^+]$) between processes. It has been shown, indeed, that local variation of the pH value (i.e. $[H^+]$) during essentially anodic oxidation of Si might lead to pattern formation in space [46]. If we look at pore etching and only consider the processes at the tip, we have additional constraints and couplings: the current at any depth of a pore with a constant diameter must be the same and must be carried by ions as a mixture of electrical diffusion currents and field currents. This necessitates both gradients for the diffusive part and potential gradients for the field part, which cannot be independent, however. On top we have particle diffusion currents for the uncharged species (e.g. HF and the reaction product H_2SiF_6 and constraints by the pore shape.

The total system thus is complex and at present beyond full modeling. Moreover, the in-situ data provided by multi-mode FFT IS still present only averages over a large number of pores. If, for example, the current inside the pore would oscillate in time (like the current in the electropolishing part of the *IV* characteristics, cf. Section 2.2.2) the average over a large number of "oscillating" pores with random phases would yield a constant average current value with little noise. This is not just an academic possibility but very real in e.g. the case of pore etching in InP [47] and also of some importance for what follows.

2.7 Beyond the SCR Model

2.7.1 The Limits of the SCR Model

The simple SCR model as presented in Section 2.5 has long since been augmented by various kinds of additions, mostly qualitative in reasoning and partially quantitative by experimentally established relations. A point in case are the always detrimental pore wall currents already mentioned (and quite visible in Figure 2.8, where the pore diameter increases towards the pore tip), which can be minimized by adding suitable surfactants, by lowering the temperature, or by otherwise modifying the electrolyte composition. Temperature is also an important factor in this context, lowering the temperature might be a crucial factor if deep pores (meaning pores with depth > 300 μm at 5 wt. % HF) are required and it is not uncustomary to systematically lower the temperature during n-macro(aqu, bsi) etching runs.

Ohmic losses in the pore, lowering the potential at the pore tip, and unavoidable concentration gradients are routinely compensated for to some extent by lowering the current density as a function of pore depth while increasing the potential somewhat. All these (and other) measures (often unpublished because of proprietary know-how) are not in conflict with the SCR model. This may appear to be no big surprise because the focusing action of the bend space charge region simply exists. However, since the introduction of the SCR model (by V. Lehmann and one of the authors (HF)) around 1990), a host of experimental findings cannot be as easily reconciled with the SCR model as the ones mentioned above.

First of all, if the pore tip is covered with oxide most of the time as stated in Section 2.5, a relevant part of the applied potential must drop across the oxide, weakening the SCR and thus any focusing action. The second basic assumption was that the current density relative to the pore cross sectional area is always j_{PSL}. If one looks at the current density per area increment at some point (x, y) and the depth $z(x, y)$ given by the shape of the pore tip, one realizes that the local current density for some area increment at the pore tip is a function (x, y) that goes through a maximum at the pore center and is zero at the rim; cf. Figure 2.10. In other words, j_{PSL} would not determine the current density actually flowing through most parts of the tip area, as already realized by Lehmann himself [1]. On the one hand this puts the justification of the Lehmann formula (Equation 2.13) on unsure footing, on the other hand, since it works so well in many circumstances, one is motivated to look at an alternative way of justifying it. The "hammer model" introduced in [3] is one example for how this could be done; Figure 2.10b gives some schematic illustration. If one were to remove periodically a layer of constant thickness (like with the bangs of an appropriately shaped hammer) or more prosaically, by quickly forming an oxide of constant thickness that then dissolve slowly, the current density would be constant and thus could be j_{PSL} throughout. If the thickness of the layers "knocked off" would be small in comparison to the pore diameter, the resulting striations of the pore walls would be small and not noticeable in normal SEM pictures. With an AFM they would be noticeable and the results of Foca et al. in this respect actually do show some kind of striated pattern [31]. In conclusion, it is entirely possible that the Lehmann equation is actually correct, but for completely different reasons as originally envisaged.

While the discussion above is more biased towards theory, we will now consider straight experimental facts that unambiguously indicate that the SCR model can at best be a part of a full macropore-etching model. The major points to consider are:

i) Diameter modulations via current modulations,
ii) Etching of n-macro(org, bsi) pores,
iii) Orientation dependence of pore etching, and
iv) New growth modes of n-macro(aqu, bsi) pores discovered recently by optimizing the (still aqueous) electrolyte not compatible with the SCR model.

The latter point will lead over to Section 2.8 where we will discuss new developments with bearing on the topic of this book.

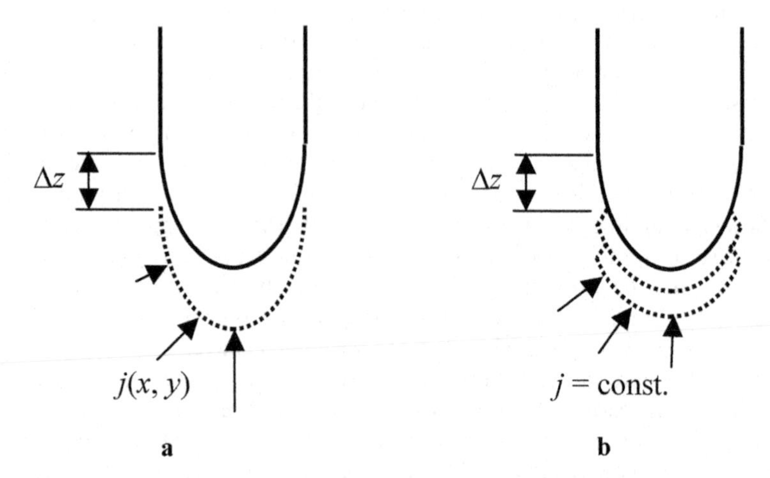

Figure 2.10. Increasing the depth of a macropore with typical tip geometry by Δz either as a smooth process in time (**a**), or layer-by-layer (**b**)

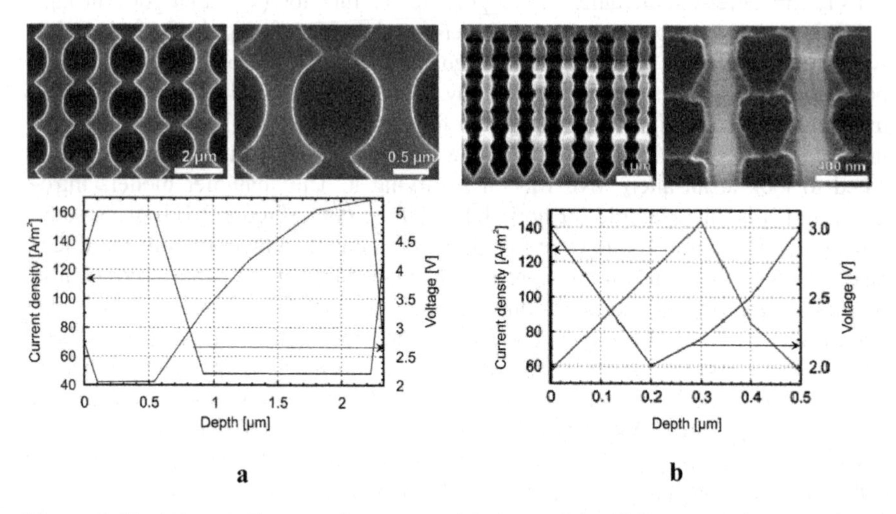

Figure 2.11. Attempts for pore diameter modulations of the Halle group. Cross sectional pictures are given together with the optimized current/potential profiles. For details see text. Courtesy of A. Langner

Modulating the diameters of macropores (d-mod for short) is important for a number of possible applications including the omnidirectional filters discussed in Chapter 11, but also photonic crystals, and Bragg or Rugate type filters, not to mention scientific knowledge. The "Halle group" (researchers at the Max-Planck-Institute for Microstructure Physics under the guidance of U. Gösele) has spent many year perfecting n-macro(aqu, bsi, d-mod) pores; the best that they (or any-body else at this point in time) can do is shown in Figure 2.11.

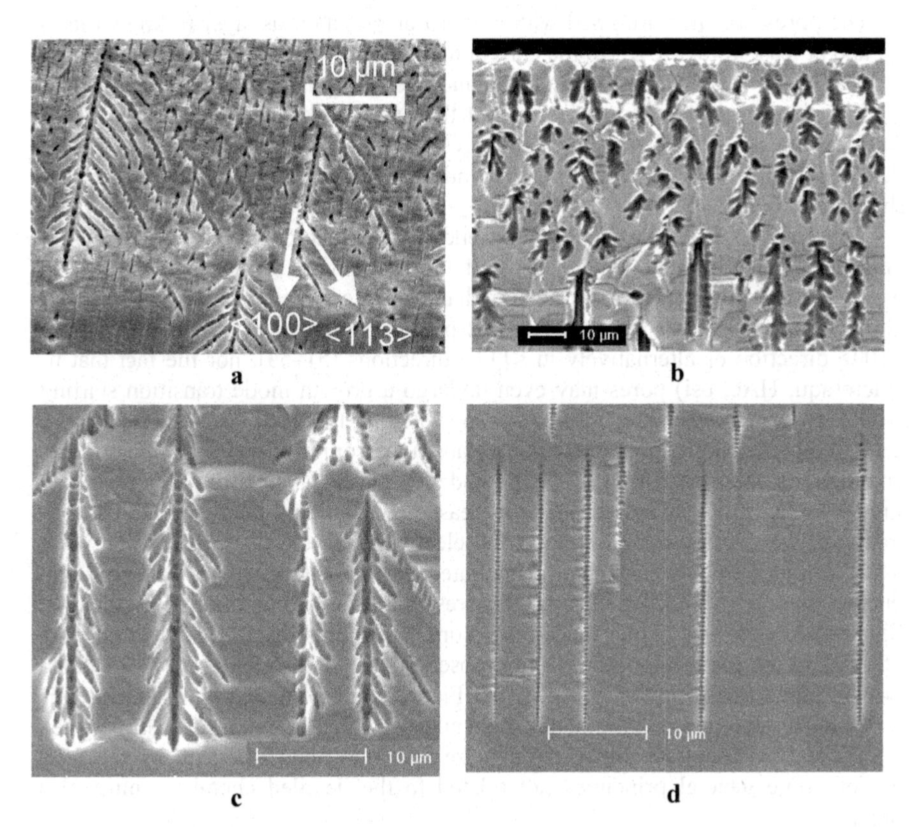

Figure 2.12. Ill-behaved macropores, not expected in the SCR model, obtained under back side illumination in organic electrolytes (**a**), (**c**) and (**d**) and, for comparison, in a high-concentration HF–HAc electrolyte (**b**). For details see text

The data shown in Figure 2.11b resulted from an attempt to reproduce the structures shown in Fig 2.11a [48] or pore geometry smaller by about a factor of 2. While the general achievement of Langner [49] is remarkable, the results cannot be understood in the context of the SCR model, which predicts unambiguously that the diameter should follow the current. Some deviations from this might be understood by considering diffusion limitations and other additions (cf. [50], but this will still fall short of understanding what is going on for the following reasons:

i) In order to obtain a roughly sinusoidal modulation of the pore diameter, it is not only necessary to modulate the current in a rather peculiar way, it also needs to modulate the applied potential in a certain way, as seen in both data sets.
ii) The optimization found for some geometry does not necessarily work for a somewhat different geometry – the $I(t)$ and $V(t)$ curves in the two cases are rather different.
iii) There are no clear rules of how to get from one case to the other one – optimization must be done by trial and error.

iv) The pores obtained are lined with a micro or mesoporous layer (a known fact) and at least in Figure 2.11b the current modulation seems to modulate the thickness of this layer far more than the diameter of the pore itself.

The pore structures shown is the best that can be obtained after > 5 years of working on the topic. Far more bizarre results can be obtained if the process is outside of the rather narrow process window found for "good" modulations as shown in Figure 2.11.

With respect to the orientation dependence of macropore growth, the SCR model remains completely quiet; at best the orientation dependence of j_{PSL} [1] would induce some variation of the pore diameter as a function of the crystal orientation. It can neither explain the fact that most Si macropores grow preferably in <100> direction or alternatively in <113> directions [51–53], nor the fact that n-macro(aqu, HAc, bsi) pores may even undergo a growth mode transition starting with growth in <113> directions and then switching to <100> as shown in Figure 2.13. Note that there is a continuous transition from well-behaved macropores completely contained within the SCR model to the rather bizarre pores shown in Figure 2.13a, if the current density is increased concomitantly with the HF concentration (always using added HAc in the electrolyte). Note also that at this point many similarities between hitherto unrelated pore structures in various semiconductors emerge. On the one hand, the pores found under conditions like those in Figure 2.13a are not unlike the n-macro(org, bsi) pores discussed above; on the other hand the growth mode transition observed is closely related to well-studied growth mode transitions of pores in InP; Figure 2.13b provided an example. This may indicate that beyond the detailed chemistry (which is quite different between e.g. Si and InP) that obviously is far more important than indicated by the SCR model, some general principles not related to the detailed chemistry must also obtain.

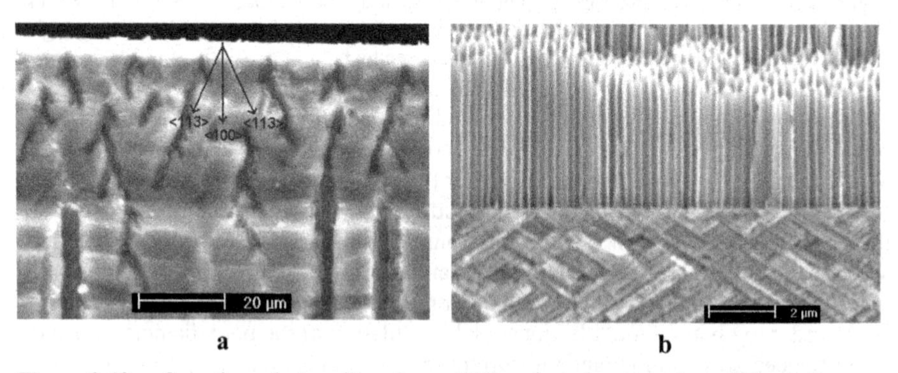

Figure 2.13. **a** Growth mode transition from <113> oriented n-macro(aqu, HAc, bsi) pores to <100> oriented pores in {100} observed at high HF concentrations. **b** Growth mode transition between differently oriented pores in InP for comparison [4]

2.7.2 Enlarging the Parameter Space for Etching Macropores

At this point we already have left the confines of the SCR model and move to very recent results that have direct bearing on etching good macropores in an economic and well-controlled way. In essence, aqueous electrolytes with either additions of acetic acid (HAc), viscosity enhancing salts (as in Figure 2.9), or both are now being used. These (aqu, HAc, visc) electrolytes induce dramatic changes in n-macro(aqu, HAc, bsi) pore etching; in particular they allow for far larger etching speeds and maximum depths with still good quality in comparison to what we now will call "neat" aqueous electrolytes. Figure 2.14. shows the former state-of-the-art with respect to pore depth vs. time. In standard 5 wt. % neat electrolytes the pore depth is typically limited to l_{Pore} < 500 μm and etching times easily exceed 12 hours. Note that curve A in Figure 2.14 already contains numerous optimization procedures not disclosed in detail here.

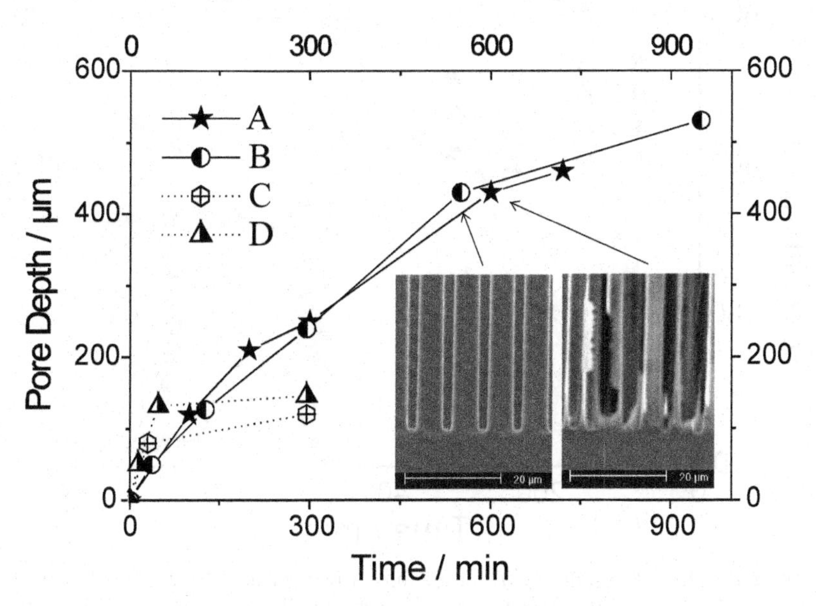

Figure 2.14. Comparison of viscous and neat 5 wt. % and 10 wt. % electrolytes with $l_{Pore}(t)$ curves; at T = 20 °C. Curve A: 5 wt. % HF (aqu); curve B: 5 wt. % HF (aqu + 0.83 g / l CMC); curve C: 10 wt. % HF (aqu); curve D: 10 wt. % HF (aqu + 0.83g / l CMC). Insets: SEM images of pore tips for the experiments marked by the arrows

The obvious remedy for faster etching is to increase the current density j. In order to keep j_{PSL}/j and thus the diameter constant, this means that j_{PSL} has to be increased via increasing the HF concentration. Of course, the illumination intensity $P(t)$ has to be increased concomitantly, which might put quite a strain on the illumination module. What happens with neat electrolyte is shown by curve C for a HF concentration of 10 wt%: the etching speed initially indeed increases by a factor of about 2, but pore growth terminates early around 150 μm. This is a quite general finding, cf. [1] and still not in direct conflict with the SCR model. Neither is the fact that making the electrolyte viscous (curves B and D) does not apprecia-bly change the respective curves, even so the pore quality (especially the pore wall roughness, cf. [31]) gets much better. One might wonder why a decrease of all diffusion coefficients in the liquid apparently does not affect the dissolution kinet-ics, but that is outside the SCR model proper.

The effect of adding HAc is not contained in the SCR model either (which ef-fectively reduces the complete chemistry to one number: the value of j_{PSL}) but rather dramatic as can be seen in Figure 2.15. The restrictions with respect to the maximum pore depth as discussed above are no longer valid, pores can now be grown at twice the speed and to far larger depths (the data shown in Figure 2.15 do not go beyond 500 µm because that was the thickness of the samples); Figure 2.16 illustrates this quite convincingly.

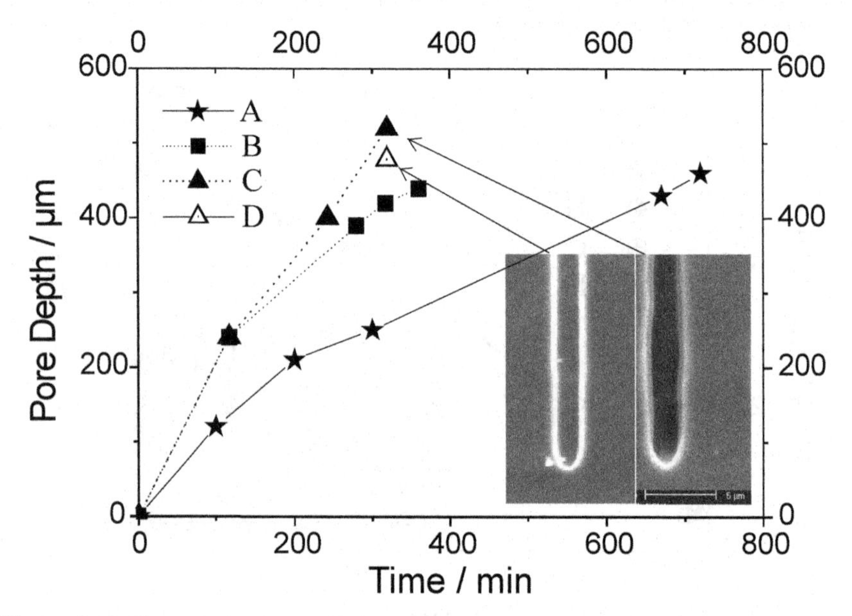

Figure 2.15. The pore depth z_{pore} (t) for the three 10 wt. % electrolytes. Curve A: 5 wt. % HF (aqu), 5 Ωcm; Curve B: 10 wt. % HF + 33 wt.% HAc + CMC, 5 Ωcm; Curve C: 10 wt. % HF + 33 wt. % HAc, 30 Ωcm; Point D: 10 wt. % HF + 33 wt. % HAc + CMC, 30 Ωcm. Insets: SEM images of pore tips for the experiments marked by the arrows

What is apparent from the figures mentioned above is that adding HAc plus CMC to the electrolyte allows for higher HF concentrations, current densities and thus etching speeds without apparent penalties. Not only is the etching speed larger at the beginning of the process, the $l_{Pore}(t)$ curves also remain linear to large depths. It is possible to increase the HF concentration and thus the etching speed even more without loosing pore depth but the reader is referred to [30] for more data as far as they exist at present. It goes without saying that the possibility to etch n-macro(aqu-opt, bsi) pores now much faster and with better quality is of great importance to applications in general and as presented here, since etching times for on wafer in the order of 10 hours are a large cost factor in any manufacturing process.

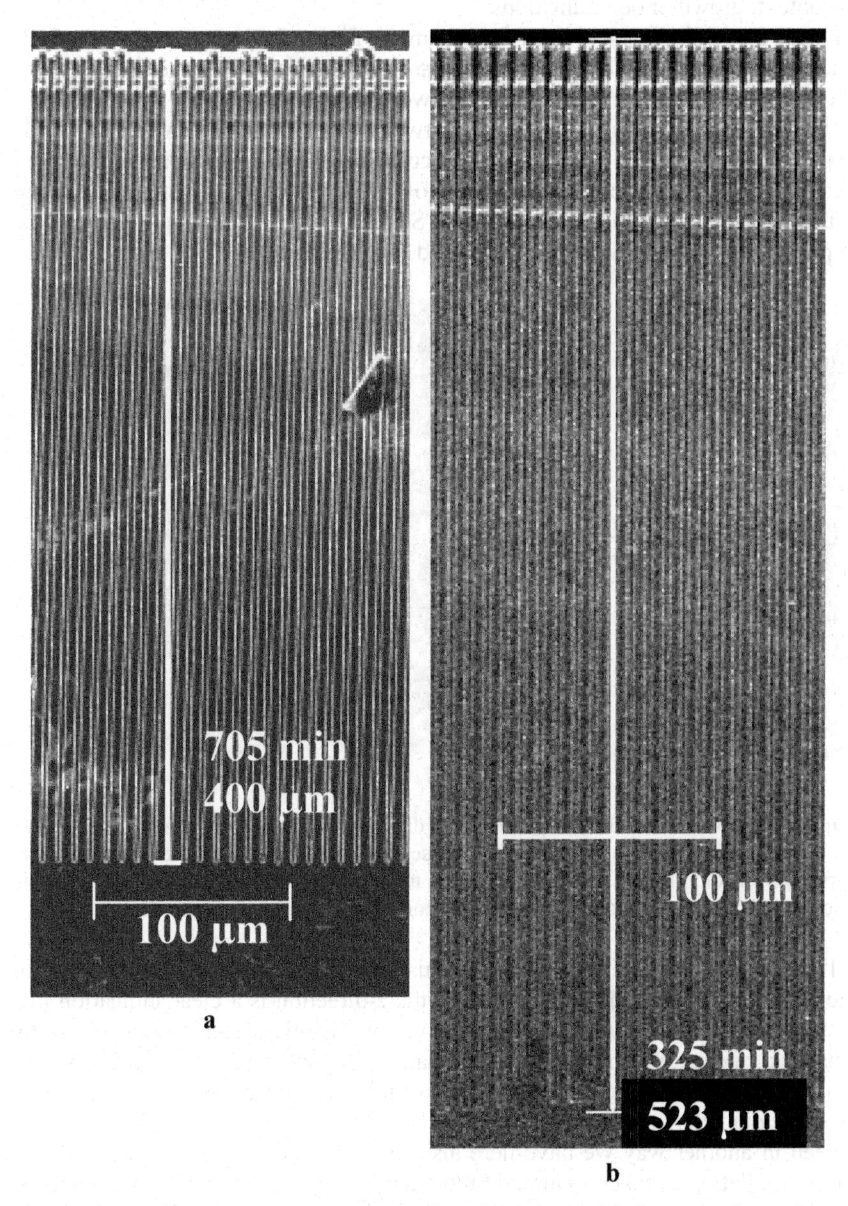

Figure 2.16. Comparison of deep, good-quality pores obtained in a standard neat 5 wt. % electrolyte (**a**) and in a 10 wt. % HF + 33 wt. % HAc electrolyte (**b**)

The few results shown here clearly indicate that the parameter space for etching good n-macro(opt-aqu, bsi) pores is far larger than believed so far. Probing this space in detail (taking into account also the dependence on doping level, temperature etc.) will go far beyond the spatial and temporal confines of this book and we will restrict ourselves to one new aspect of macropore etching that has emerged in this context: growth mode transitions.

Figure 2.13a already showed a first example, Figure 2.17 gives one more. Self-induced anti-phase diameter oscillations are often observed in the high [HF] – CMC region of parameter space. In other words, the pore diameter starts to oscillate but with an anti-phase correlation between (some) neighboring pores: while one pore becomes wider, its neighbor reduces. Note that for the hexagonal array of pores as present in Figure 2.17, a top view of the pores shows a frustrated structure (Figure 2.17c) i.e. no apparent order (cf. Section 3.4): a "thick" pore cannot be completely surrounded by six thin pores and vice versa.

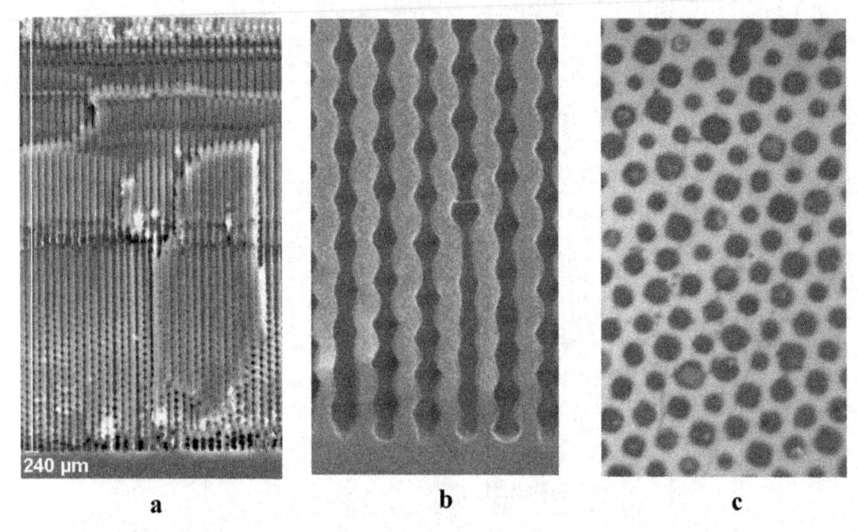

Figure 2.17. **a** Self-induced "anti-phase" pore diameter oscillations in n-macro(aqu, HAc, visc, bsi, hex-litho) pores. **b** Details of anti-phase oscillations from a different sample under comparable conditions. **c** Top view on a cut made in the anti-phase oscillation region. "Thick" and "thin" pores show a frustrated structure without any apparent order

The fact that the pore arrays even in this case obviously "feel" the need to induce some additional ordered structure in the *z*-direction is a clear indication that some mode transitions as generally observed in chaotic systems upon smoothly changing a relevant parameter has taken place. What we have here inside a pore is a self-induced current oscillation in time, alluded to already in Section 2.2, that causes the diameter oscillations.

Seen in another way we have here also the second example of a self-induced current oscillation in space as already illustrated in Figure 2.4 that is superimposed on the one externally determined from the lithographic structuring. One might speculate that the lithographically enforced initial current oscillation in space expressed in the hexagonal pore "single crystal" is no longer within the bandwidth of an internal length scale and that the system now starts to express its own spatial frequency scale.

Why does the addition of some acetic acid (or other carbonic acids) allow to grow good n-macro(aqu, bsi) pores with parameters not achievable with classic electrolytes? The SCR model clearly cannot say much to this finding. The basic explanation suggesting itself is that pore wall passivation is better in this case, reducing the pore wall currents I_{PW} mentioned in Section 2.5. If we assume the pore wall current density j_{PW} is constant, I_{PW} must increase linearly with pore length l_{pore}. An ever-increasing part of the external current I is thus flowing through the pore walls, leading not only to an increase of the pore diameter and a concomitantly conical shape, but also to a decreasing pore tip current. The pore, in other words starts to grow more and more in width and less in depth. Pore wall currents are thus always destructive: they modify the pore geometry in unwanted ways and ultimately limit the achievable pore depth. Raising the external potential and current in order to supply the pore tip with what it needs for steady growth in some large depth unfortunately tends to strongly increase the pore wall currents, so nothing is gained by doing this, it rather makes things worse. Lowering the temperature may help to some degree because this may decrease pore wall currents without interfering too much with the dissolution process at the pore tip.

Of course, the above consideration is hypothetical. It is an interpretation of many experimental findings and not (yet) based on in-situ data about pore wall currents and the like. In-situ dual-mode FFT IS might change this; but the technique is new and specific results to this point must still be awaited.

What determines the magnitude of pore wall currents? The first factor is the concentration of available holes; this is expressed indirectly in Equation 2.4 that gives something like the maximum pore wall current over a simple junction. The second and more important factor for Si (but less so for Ge, cf. [33]) is what was termed "pore wall passivation" above, and this qualitative term denotes everything that makes the flow of carriers across the solid liquid interface difficult. For example, hydrogen passivation, often referred to in the context of Si electrochemistry, would charge the surface positively (via $Si\text{-}H^+$ formation) and thus impede the flow of holes in the surface direction. A thin oxide layer (or any other insulting layer) would completely impede current flow or at least make it more difficult. The "right" surfactant might be helpful, or the "right" electrolyte chemistry in the sense of the additions noted in Table 2.2. The physically inclined reader will now associate "pore wall passivation" with "black art" and it cannot be denied that this is a viable point of view at this point in time.

In a general sense, pore formation can quite generally occur if there is a gradient in the degree of passivation between the pore tip and the pore walls. Eschewing the pore nucleation mechanisms, growth of a pore will simply occur if more current flows through the pore tip because it is less well passivated than the pore walls. Concentrating more holes around the pore tip area, either by using the focusing effect of the SCR or by inducing some localized breakdown, will help, but is not necessarily instrumental anymore. In what follows this point will be considered in some detail in the context of the "current burst model".

2.7.3 The Current Burst Model

From the aforesaid it has become clear that the SCR model, though very useful in some part of parameter space, falls far short of covering major aspects of just n-macro(bsi) pores, not to mention other pore types like other kinds of macropores (see Section 2.8), mesopores and micropores. This is not to say that it is wrong; the existence of a (bend) space charge region of a certain thickness d_{SCR} is certainly one of the most important ingredients for most types of pores. It has also become clear that self-organization features are encountered in pore etching, expressing themselves in distinctive pattern formations in space and time or slightly less spectacular in self-induced growth mode transitions. There are far more observations of these two topics in general pore etching in semiconductors, and in all cases of an observed self-organization one must assume that the underlying pore etching system is "chaotic" in the sense that non-linearity together with feedback allows various types of order and chaos plus sudden transitions between such modes. Since self-organization is not always observed in pore etching experiments, one might ask if a "chaotic system" is the rule or the exception. In what follows we use the term "chaotic system" loosely and only in the sense that the system is capable of self-organization under certain conditions and not in the sense that all criteria for deterministic chaos are met.

While there is no undisputed answer to this question, it is helpful to point out that lack of directly apparent self-organization does not rule out an underlying chaotic system. If etching would actually proceed in a "chaos" mode for a single pore, averaging over many pores may still produce smooth parameters. Strong self-organization, for example oscillating currents inside all pores, will also go unrecognized in external quantities if there is no phase correlation between the current in different pores. The same is true for electrochemistry outside of the pore etching region, e.g. for electropolishing of Si in aqueous HF electrolytes. While strong self-organization in the form of self-induced current (or voltage) oscillations is observed after the j_{Ox} peak (cf. Section 2.2), it has been unambiguously shown that these oscillations are already present at potentials where they are not observed in the external circuit. What happens is that only parts of the surface (so-called domains) oscillate with a certain phase, and averaging over many uncorrelated domains produces constant external parameter values [54].

It has been mentioned before that the chemical part of pore etching actually does provide for a non-linear system with many internal couplings or feedback (cf. Section 2.6.2). While this might be sufficient to supply the ingredients necessary for the "chaotic system" defined above, it is not likely. First, while there are many examples of purely "chemical" pattern formation, it is still rather the exception than the rule. Second, the self-organization features observed during electropolishing could not be explained by reaction kinetics alone. In contrast, the oscillations mentioned above have been modeled quantitatively in great detail with the so-called current burst model (CBM), which we will now introduce superficially because it also applies to pore etching, albeit qualitatively in this case. For more details about the CBM refer to [55,56].

The CBM model is conceptually simple. Its basic assumption is that current flow through a semiconductor electrode often proceeds over a barrier that is overcome in a non-linear fashion. The concept is best illustrated by looking at the electropolishing of Si, i.e. a dissolution reaction dominated by the oxide formation as described in Section 2.3.3 in Equation 2.10 (bear in mind that electropolishing also occurs to some extent at the tip of a growing macropores to appreciate the relevance to the topic of this chapter). It is an undisputed and experimentally established fact that the Si electrode is covered with a thin (several nm) layer of oxide at all times during electropolishing. As pointed out in Section 2.7.1 there is strong evidence that the macropore tip is also covered with oxide most of the time so what follows applies to some extent to pore tips, too.

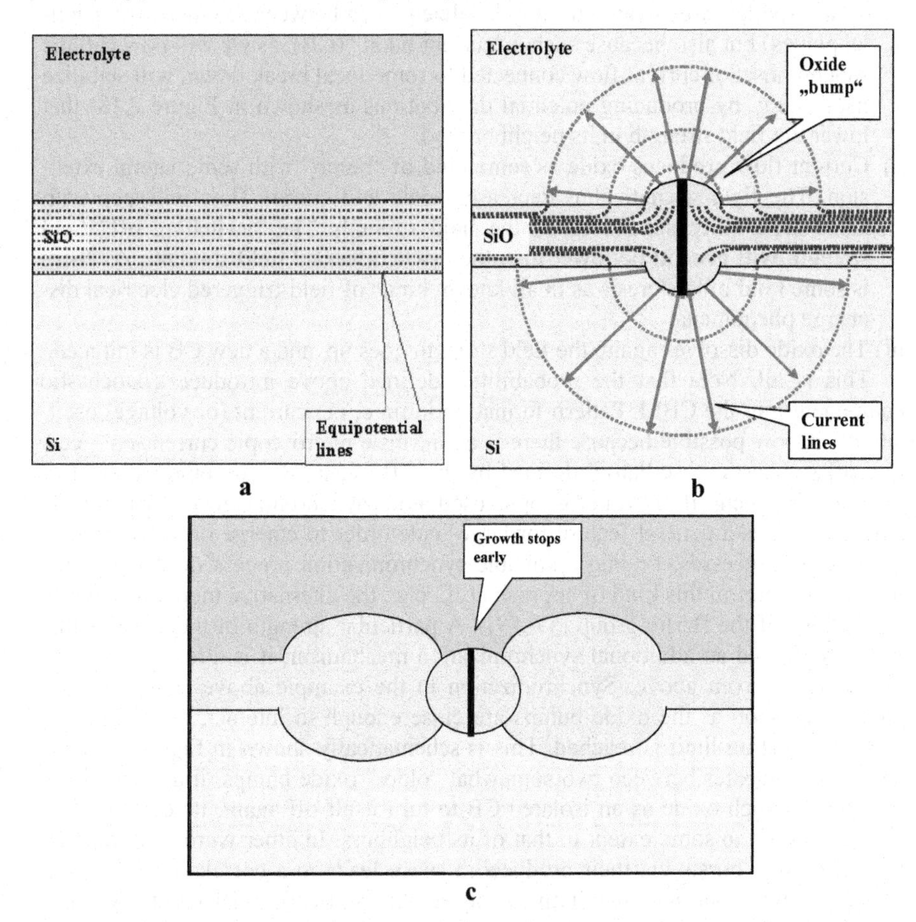

Figure 2.18. Schematic illustration of a current burst through an oxide layer. **a** Dissolving oxide layer. **b** A current burst occurs with some probability at high field strength = thin oxide, initiating rapid oxide growth and a concomitant redistribution of current and field lines. **c** A current burst initiated between "old" oxide bumpy will stop earlier

If, for ease of conception, we start potentiostatically with a relatively thick SiO_2 layer as shown in Figure 2.18a, the applied voltage will completely drop in the SiO_2 with equipotential lines as shown. In an HF bearing electrolyte the SiO_2 will dissolve and the field strength inside the oxide increases. For a typical voltage of 5 V and a thickness of 5 nm the nominal field strength is 10 MV/cm and this is around the known breakdown field strength of good SiO_2. Electrical breakdown will occur at some critical field strength inducing current flow at a still finite oxide thickness. That this happens is beyond doubt and not yet a critical assumption for the CBM. The critical assumptions are simply:

i) Breakdown will not happen globally but locally – in small nm sized regions- with some **probability** depending on the local field strength. This is not only a rather general truth for all kinds of electrical breakdowns (flashes of lightning or all "sparks" are always strongly localized, even between tow perfectly parallel planes) but also because such a "current burst" (CB), as we will now call the sudden onset of current flow connected to some local break down, will stabilize itself – e.g. by producing potential distributions as shown in Figure 2.18) that lower the field strength in its neighborhood.
ii) Current flow produces oxide as some kind of "bump" with some lateral extension. The field strength thus decreases again and current flow will stop with some probability depending on the field strength. The "switching off" field strength will general be lower than the "switching on" field strength, i.e. there is some kind of hysteresis as in all known kinds of field-triggered electrical discharge phenomena.
iii) The oxide dissolves again, the field strength goes up, and a new CB is initiated.

This is all. Note that the probabilities defined above introduce a stochastic component into the CBM. Pattern formation in time, i.e. current (or voltage) oscillations, is now possible because there is an intrinsic microscopic current-on – current-off stochastic "oscillator" defined by the CB cycle. Macroscopic pattern formation will occur if the microscopic oscillators synchronize to an appreciable amount. This is a general feature for large-scale order to emerge out of stochastic small-scale processes; finding a suitable synchronization process needed is often the difficult part in this kind of approach (cf., e.g., the alternative model for current oscillations of the Berlin group [57–59]). A particular strength of the CBM is that it does not need an additional synchronization mechanism; it is already contained in the items given above. Synchronization in the example above "automatically happens as soon as the oxide bumps are close enough to interact, i.e. as soon as some percolation limit is reached. This is schematically shown in Figure 2.18c. If some CB nucleates between two somewhat "older" oxide bumps, it does not have to grow as much oxide as an isolated CB to turn itself off again. Its cycle time is now correlated to some extent to that of its neighbors. In other words: An interaction of current bursts via their products in space leads to a correlation of CBs in time. As it turns out in a quantitative analysis, this is already sufficient to synchronize CBs and to model the current/voltage oscillations in the electropolishing region quantitatively in great detail [60].

Turning to pore growth now, the CBM implies rather directly that current flow through the oxide at a growing pore tip follows similar principles. The fact that good n-macro(bsi) pores only grow if the valence is around 2.7 indicated that the Si dissolves by direct dissolution and by oxidation in about equal amounts, but that does not imply that the pore tip is not covered with oxide most of the time because direct dissolution is a fast process, taking out its part of Si from some "pixel" in short time t_d while the oxidation part of the same pixel is slow and will have the pixel covered with oxide for a long time $t_{Ox} >> t_d$.

However, the CBM model is easily extended to direct dissolution, too, In this case the barrier layer is not an oxide but whatever else passivates the electrolyte – Si interface. For the process in question this will be the well-known hydrogen coverage in acidic electrolytes [1,2,61]. The mechanism of passivation in this case is simple to understand. Hydrogen, by forming Si-H bonds with surface atoms, removes interface states in the band gap of the Si that pin the Fermi energy and thus prevent the formation of a substantial space charge region in the Si. Reducing the density of surface states increases the SCR over which some part of the applied voltage drops. Fully passivated interfaces in the sense of allowing sizeable SCRs will drop a larger part of the applied potential across the SCR than badly passivated ones. This implies that the interface potential actually driving the chemical interface ratios will be lower for passivated surfaces, drawing only small currents.

Passivating a freshly produced interface will take some time, however, and the relevant time constants depend also on the crystallographic orientation of the exposed surface. This is directly felt in some pore etching cases (most prominently InP) where short stops of some pore etching process (i.e. switching off the current for half a second) may not be felt at all, while longer breaks (a few seconds) force the system to "start all over again" with pore structures completely different from the ones obtained before the stop. The CBM thus "focuses" current at the less well passivated parts of a surface, and that is exactly the point that was made above at the end of 7.2 in order to give an explanation for pore formation in general.

A more detailed analysis can be found in [4,62] and will lead to the "hammer" model alluded to before in Section 2.7.1. For the purpose of this section we will only focus on one crucial point in this context: If, after some cycle involving a current burst, the interface is locally left without current flow for some time for whatever reason, it will start to passivate. Nucleating a new CB at this location will get increasingly more difficult with time then and this introduces quite generally an interaction in time between CBs. In other words, the probability of initiating a CB at location (x, y, t) is not only a function of the field strength at $\{x, y, t)$, but also of the time Δt that has elapsed since the last current burst at (x, y). As we will see, this is already a sufficient condition for forming pores.

This rather simplified and supercilious description of the CBM allows drawing a number of conclusions with respect to pore etching in Si. First we will give an alternative one of Lehmann's equation. All we have to do is to assign an average current density to a CB. If we now look at the basic *IV*-characteristics for p-type Si in an aqueous HF electrolyte, raising the current density with increasing potential simply means that the density of (randomly distributed) isolated CBs is raised until at j_{PSL} the maximum CB density and thus also current density is reached. The Lehmann equation follows if we now assume that macropore tips can only grow coherently if the tip is always covered with oxide demanding close packing of CBs and thus j_{PSL}. The dissolution mechanism is quite different, however, to the ideas expressed in the original SCR model as shown in Figure 2.10.

2.8 Micropores, Mesopores and Special Macropores

2.8.1 Micropores and p-Macro(aqu) Pores

One might ask why a dense layer of CBs right at the j_{PSL} as suggested above does not induce current oscillations, and the answer is that in the part of the characteristics just before the j_{PSL} peak microporous Si is formed. While there is no detailed model of micropore formation, it is generally accepted that the "quantum wire" or better "quantum dot" nature of the Si between micropores with its concomitant increase of the band gap prevents holes from moving into the walls and thus prevents dissolution of the walls. It can be doubted, however, if this quantum wire effect is sufficient to explain micropore formation because it is a general semiconductor effect – but no micropores have ever been found in, e.g., III-V semiconductors. The CBM with a negative correlation in time between CBs plus "quantum dot behavior is able to produce the required "quantum sponge" in principle, and this will be the second example.

Imagine one CB on a bare Si surface with the dissolution process dominated by direct dissolution but with some oxidation as well. The time sequence must proceed from mainly direct dissolution to oxidation because the oxide formed will eventually switch off the CB. The direct dissolution part of the CB will dig a small micropore into the Si, the oxidation part will pluck this pore with an oxide lump. Under galvanostatic conditions a constant current must flow, resulting in complete coverage of the surface with CB-induced small holes rather soon for current densities not much smaller than j_{PSL}. As soon as distances between pores become smaller than a few nm the quantum wire effect prevents further dissolution (no more holes). Current must now flow through the already produced pores, i.e. new CBs must nucleate somewhere down the pore. There is now a "negative interaction" in time: places that have harbored a CB a time t, are less likely to produce a new one at time $t + \Delta t$. If the oxide pluck has not yet dissolved, the probability of nucleating a CB at the tip of the pore produced by the former is low and the new one starts somewhere else inside the pore in some other direction. Quantum dot effects are thus necessary but not sufficient to produce a microporous nanosponge; certain oxide formation and dissolution properties are needed, too.

As a last example that will bring us the next section concerning p-macro(org) pores, we consider the case of negligible oxidation that is prevalent, for example at current densities well below j_{PSL} for p-type Si in aqueous electrolytes or for electrolytes with little "oxidizing power" (cf. Section 2.4.1). i.e. organic electrolytes. A freshly etched part of a Si surface where current flow has temporarily stopped with the termination of some CB cycle will start to passivate, making the nucleation of a new CB more difficult with time. Galvanostatic conditions, however, enforce a constant (average) nucleation rate. What is going to happen, starting with a random distribution of widely spaced CBs matched to the low current density in relation to j_{PSL} is that CBs will start to cluster, forming close-packed domains because now there is a positive interaction in time: places that have harbored a CB a time t, are more likely to produce a new one at time $t + \Delta t$ since as longs as that place did not yet passivate completely during Δt. Since current flow etches Si, a domain will eventually trap itself locally by growing a pore. The prediction is thus that under these conditions macropores will form and that their nucleation might take some time. This is exactly what happens, cf. [36].

The p-macropores(aqu) described here were first described by Lehmann [63]; an example is given in Figure 2.20b. However, microporous Si is usually produced with electrolytes containing a large concentration of HF (often just 49 % HF mixed with an equal volume of ethanol), and at current densities close to the j_{PSL} value of the system, while p-macropores(aqu) are found at current densities much lower than j_{PSL} and for medium to low HF concentrations.

We will not belabor this point here anymore because p-macro(aqu) pores have not been of much value so far. However, to demonstrate that the CBM goes much farther than is apparent in the brief summary given her, Figure 2.19 shows a semi-quantitative prediction about the areas in parameter space occupied by p-macro(aqu) and p-micro(aqu) pores.

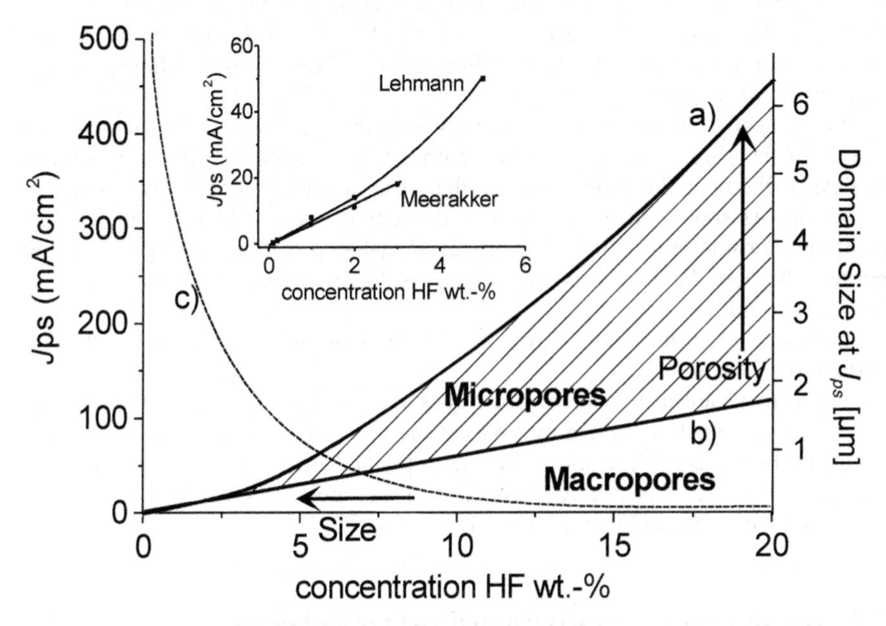

Figure 2.19. Semiquantitative predictions from the CBM (after [3])

A last point worthwhile to mention is that understanding the rather non-linear if not to say bizarre behavior of many pores with respect to current modulations, i.e. with respect to their response to an extrinsically impressed time constant, demands some intrinsic time constants at the minimum. The CBM has an intrinsic time constant – the cycle time of a CB. This is not to say that the CBM explains this behavior but only that in contrast to the SCR model it is complex enough to potentially account for those phenomena.

2.8.2 Mesopores

While macropore etching proved to be rather complex and sensitively influenced by a large number of parameter, the opposite is true for mesopores as long as this term is used for all pores essentially produced by a local breakdown of the SCR. As pointer out in Section 2.5, this always happens for practically all applied potentials if highly doped Si (or any other highly doped semiconductor for that matter) is used. The pore diameters will be defined by the critical radii needed for breakdown, the distance between pores by the width of the space charge region. Since these two length scales are comparable, quite often pores with "meso" dimensions result, sometimes also nominally with "macro" dimension, but the term "mesopore" is usually still applied.

The rather simple formation mechanism makes the etching of mesopores –n,p-meso(breakdown) pores in the nomenclature used here – far simple than macropores. No illumination is needed, nucleation and formation tend to be rather uniform even in non-optimized simple etching cells, and the porosity follows the current density over a wide region.

The typically "fuzzy" morphology of mesopores is a result of the inherent stochastic of the pore formation process. With highly non-linear "run-away" effects like avalanche breakdown or tunneling, there can be no quiescent "steady state" at a pore tip. As soon as a locally high field strength starts breakdown, the rapidly increasing local current will lead to processes that stop current flow again – either because the pore diameter becomes to large to large, or some oxide is formed inhibiting current flow.

In other words, mesopores are naturally following some "current burst" scenario as outlined above.

While there are a number of open questions concerning details of mesopore formation, not much more can be states at the moment than what has been established in the remarkable work of Lehmann [8] and in [1]. For some equally remarkable structures that can be obtained with mesopores, the reader is referred to the work of, e.g., M. Sailors group [64,65] or [66,67].

2.9 Macropores in Lightly Doped p-Type Silicon

2.9.1 The Basics

The first p-macro(org) pores were found by Propst and Kohl in 1993 [19,68] with an electrolyte containing almost no water (and thus with small oxidation power as noted in Section 2.4.2) They came as a big surprise, because they seemed to be clearly at variants with the then prevailing SCR model. In the following years, many papers have been published dealing with these new kinds of p-macropores(org) obtained by using different kinds of organic electrolytes, cf. [20,51,63,68]. The preferred types of organic solvents then were Dimethylformamide (DMF) and Dimethylsulfoxide (DMSO). While it first appeared that p-macropores(org) were limited to Si with a rather high resistivity of $\geq$ 100 Ωcm (resulting in relatively short and "bulgy" pores), Ponomarev et al. [20] were the first to etch macropores on 1 Ωcm (100) and (111) orientated Si substrates using different organic electrolytes. Christophersen et al. finally found very stable growth conditions for p-type macropores (org.) allowing for pore depths up to 400 μm [51,69] and the work of Bergstrom's group has pushed the art of etching p-macro(org) pores even farther [70,71].

In the review [3] macropores in p-type Si were divided into p-macro(aqu) and p-macro(org) types, depending if they were obtained in aqueous electrolytes at current densities well below j_{PSL} or in organic electrolytes with minimized water content. Meanwhile this distinction may appear to be no longer very useful because very good, if not the best p-macropores with depths exceeding 500 μm, are obtained, for example, in HF (49 wt%) : EtOH : H_2O = 1:2:3 electrolytes by P.L. Bergstrom's group in Michigan farther [70,71] and thus in electrolytes that are not exclusively "organic" or "aqueous" in the sense defined in [3]. However, it will be shown later that there are good reasons to address these pores as a new class outside of the p-macro(aqu, crysto) or p-macro(org, crysto) pores, which are always crystallographic ("crysto") in nature. One might argue that the p-macropore situation has rather become more puzzling with time, and while this is true, it is also true that a better understanding is also emerging and that this might be decisive for making p-macropores useful in the future.

We may consider three deep questions in connection with p-macro(aqu; org) pores – one more fundamental, one more practical, and one in between:

i) What is the formation mechanism? Which parameters determine geometry and morphology and what is the role of the electrolyte composition?
ii) What is the bandwidth of achievable macropore geometry and how can it be controlled? In other words: given the inexhaustible (chemical) parameter space, how do we find optimized etching conditions for a particular task, i.e. electrolyte composition, current density, voltage, and temperature?
iii) How about systematic pore diameter modulations as need for some (optical) applications or as "probe" for formation theories?

2.9.2 Some Specifics

Starting with formation mechanisms, a straight answer is that while there is no lack of models, there is very little predictive power, in particular with respect to electrolyte composition. Mostly "chemical" models [19,72,73] have been proposed, as well as a detailed analysis of the stability of a moving Si – electrolyte interface [74] with the result that it can become unstable and "decompose" into a wavy interface, i.e. into pores. Lehmann was more inclined to look at the semiconductor side, producing good arguments for the point of view that the current is still focused on pore tips because of easier hole transfer in the strong electrical field that is still present at the pore tips, even for a forwardly biased Schottky-type p-type Si – electrolyte junction. The Kiel group advanced some arguments centered around passivation of side walls in the context of the current burst model [61] as already discussed in Section 2.7.

None of the models proposed is wrong, and there might be cases where one particular mechanism might even prevail; for example pore formation in amorphous Si might indeed proceed with the "interface instability" mechanism suggested in

[74]. However, for the good and very deep macropores as shown in [70], which are of interest here, the question is not only why and how those pores form at all, but also what keeps their growth stable at very large depths exceeding 500 μm.

A basic problem in this context is that the system has only one degree of freedom. Current just flows, either according to the setting on a galvanostat at whatever voltage it takes, or with whatever current value establishes itself for potentiostatic conditions. The essential prime input parameters thus are: one electrical parameter (I or V; possibly modulated), the Si doping level (still determining the width of some SCR), and the electrolyte composition. The only secondary parameter with some possible (small) importance is the temperature. If the minority carrier lifetime τ or other semiconductor parameters are of any importance is neither likely nor known at present. The substrate orientation has a similar role as in the case of n-macro(aqu, bsi) pores [75] as long as macropores are growing into a crystallographic direction (either <100> or <113>), i.e. p-macro(org; aqu; cryst) pores.

Here we will not engage in any more speculations about the prevailing mechanisms except to note that the remarks made in connection with the current burst model in Section 2.7, as well as the reasoning behind the "oxidation power" (Section 2.4.2 and [76]) are still valid as far as they go. Looking at the example of p-macro pores in Figure 2.20 from P. Bergstrom's group (taken from [77] and probably the best that can be done at present, we rather ask: how is it possible to produce very good p-macropores with lengths exceeding 500 μm, considering that this is a difficult task for n-macro(aqu, bsi) pores? Since the latter always experience the focusing effect of the bend SCR (at least to some degree) and certainly have no more holes available for the destructive pore wall currents than p-type Si, they would be expected to grow more stable into the depth as p-macropores in any kind of model.

The tentative answer, proposed for the first time here, is that it is very likely that "good and deep" p-macropores are actually of the current-line type (see Section 2.4.1) and thus must be treated as a new pore type as indicated above. A clear indication for this is given in [71] where {111} oriented p-type Si was investigated. Crystallographic p-macro(org, cryst, <113>) pores were found at low current densities and for "true" organic electrolytes, whereas for electrolytes with more oxidation power (produced by the addition of some water) current-line pores (p-macro(org/aqu, curro) pores) resulted. Switching the current density repeatedly even allowed to produce a sequence of crysto/curro pores quite similar to the structures obtained for InP [4]. In passing we note that while current line pores have been observed in Si before [78], they are the exception and not the rule.

Comparing the conditions for the generation of the very deep p-macropores in {100} Si from [70] to the ones given in [29] makes it quite likely that these deep macropores are current-line pores, too. Since the growth direction of current-line pores is perpendicular to the surface and thus coincides with <100> for the usually used {100} substrates and p-macro(curro) pores, they are indistinguishable from p-macro(crysto, <100>) pores by their geometry alone. Impedance spectroscopy has made some contribution to the crysto/curro pore topics in InP [43, 44] and might be used for p-macropores, too, but no experiments along theses lines have been reported so far.

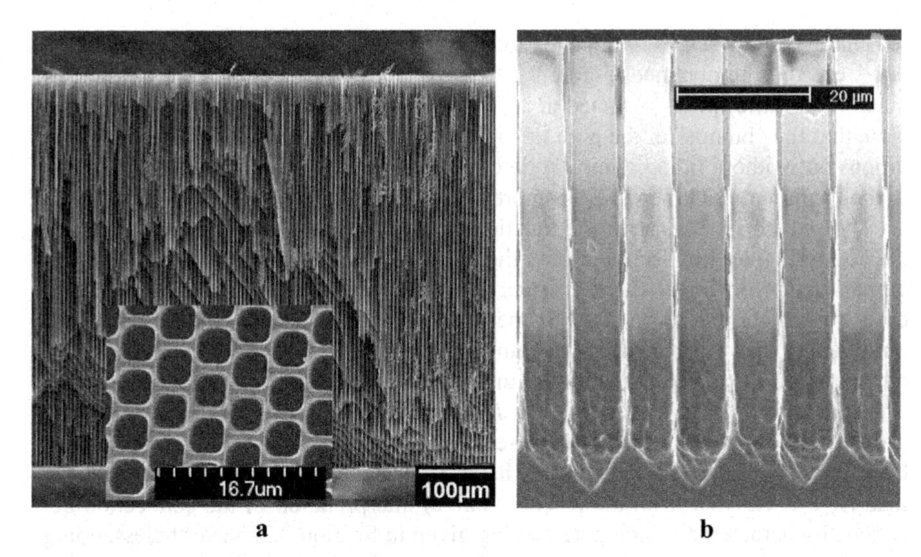

a **b**

Figure 2.20. **a** Macropores in lightly doped p-type Si as produce by P. Bergstroms group ({100} (17–23) Ωcm Si sample, HF (49 wt%) : EtOH : H_2O = 1:2:3, j = 27 mA/cm^2); the inset shows the top view. **b** Macropores obtained at very low current densities (far below the j_{PSL} peak in aqueous electrolytes (cf. [3]). **a** Courtesy of P. Bergstrom

From InP it is known that current line pores need some kind of diffusion limitation and need a more or less involved nucleation process, typically a layer of dense crystallographic pores, before they can grow. This corroborates the finding that the deep p-macropores discussed here are only found if nucleation is defined lithographically and if pore growth is started with some attention to the initial stage. From not yet published results of the Kiel group around one of the authors (H.F.) it is advantageous, for example, to start with a higher current density than the one used for stable growth.

Accepting this hypothesis changes the original question about the formation mechanism to the question about the basic difference between crystallographic pores and current-line pores. This leads straight into the InP pore etching literature plus speculation about how the findings from this material can be transferred to Si, and we will not go into this her except to refer to some InP key papers in this context [4,79–81].

The second question about the bandwidth of achievable p-macropore geometries and so on is relatively easy to answer: The basic intrinsic length scale determining pore geometry is d_{SCR}, the width of the space charge region with a direct dependence on the square root of doping concentration and potential drop at the junction. Note that a classical pn-junction would hypothetically draw a current in excess of 1000 A/cm^2 if the applied voltage equals the build-in voltage to appreciate that even a forwardly biased p-type Si/electrolyte junction has still an appreciable SCR.

The general rule with respect to the p-macropore geometry is that the remaining Si must be field-free, demanding that the distance between pores is in the order of 2 d_{SCR}. The pore diameter then simply adjusts to whatever value is needed to clear out the area between the "impenetrable" walls containing the SCR. Rather large pores can thus be obtained; Figure 2.20b gives an example. In this case we have p-macro(aqu, litho) pores obtained at $j << j_{PSL}$ in a standard aqueous HF electrolyte. Note that the "bumps" at the pore tip signal the size that pores under identical conditions but without lithographically determined nucleation would assume, and refer to the "hammer model" in [62] for more details.

The growth speed is strictly a function of the current density j_G or better j_{Pore} for the reasons given in Section 2.3.1; the question is thus what limits the useable current density? The *IV*-characteristics (always obtained with flat surfaces) in this case are not helpful. Even if a PSL peak or some other special current density is observed, it has no particular meaning since the *IV*-characteristics during pore etching are quite different and rather unknown, in particular for current line pores. The reason for this is that taking an *IV* curve during pore growth interrupts the growth process and is thus not representative. Very few data in this respect have been published. Good p-macropores have been etched at current densities larger than j_{PSL}; at variance with the (by now naive) interpretation of the connection between *IV*-characteristics and pore etching given in Section 2.2. Nevertheless, going beyond certain critical current densities does not allow pore growth anymore; some quantitative data for this can be found in [70] In retrospect, the fact that global growth rates in excess of 1 μm/min have been reported for deep p-macropores (a feat impossible for n-macro(aqu, bsi) pores until the progress reported in Section 2.7.2) is easily understood assuming that those pores were of the (always fast growing) current-line type.

The last question about systematic pore diameter modulations is difficult to answer. Besides some unpublished results from the Kiel group to this topic, nothing seems to have been published in this respect. The reason might be that attempts to modulate p-macropore diameters by modulation of the current typically produce no discernable effect. This simply means that current changes do not change the pore diameter but the growth rate. To some extent this finding is clear: The pore will always grow to a diameter that produces field-free Si between pores because that is part of its growth mechanism. It thus cannot respond to current density modulations by changing its diameter. If the modulation amplitude is large enough, it could respond by a growth mode transition, e.g. from curro-pores to crysto-pores, i.e. in a non-linear way. This is speculative however (to the point where it goes beyond what has been published in [71] and will thus not be pursued any more here.

In conclusion it can be stated that on the one hand, p-macropores have a large potential for applications where good and very deep pores are needed as long as constant diameters are good enough. On the other hand, looking at p-macropores from a more fundamental point of view makes clear that pore etching in Si is still a topic for fundamental research.

2.10 Summary and Conclusion

In the course of this chapter we have moved from some basic considerations concerning pore etching known since about 1990 to rather involved and very recent points. If a reader not previously familiar with pore etching in semiconductors feels a bit confused and overwhelmed a this point – that is just natural. Pore etching in semiconductors, as should have become clear, resides in some intersection of semiconductor physics, chemistry and electrochemistry, stochastic physics (or chemistry), analytical techniques like impedance spectroscopy and – not to forget – sophisticated hard and software engineering; it is not something easily subsumed under some simple equations or rules.

Nevertheless, certain standards and useful rules do exist, embodied, for example, in the SCR model that describes rather well some particular corner in parameter space. The fact that there are other areas in parameter space where pore formation might not be well understood, and moreover regions not yet explored, does make the picture fuzzy if colorful. This must not be seen as a detriment to the topic of this book but as a very positive potential. Considering that the present state of pore-etching art is not always evolved enough to meet the very difficult demands of optical applications, the fact that there is much room for further improvements is simply what is needed.

2.11 References

[1] Lehmann V, (2002) Electrochemistry of Silicon, Wiley-VCH, Weinheim.
[2] Zhang XG, (2001) Electrochemistry of silicon and its oxide, Kluwer Academic – Plenum Publishers, New York.
[3] Föll H, Christophersen M, Carstensen J, Hasse G, (2002) Mat. Sci. Eng. R, 39(4): 93.
[4] Föll H, Langa S, Carstensen J, Lölkes S, Christophersen M, Tiginyanu IM, (2003) Adv. Mat., 15(3): 183.
[5] Smith RL, Collins SD, (1992) J. Appl. Phys., 71(8): R1.
[6] Bisi O, Ossicini S, Pavesi L, (2000) Surface Science Reports, 38: 1.
[7] Ossicini S, Pavesi L, Priolo F, (2003) Light emitting silicon for microphotonics, Springer, Berlin.
[8] Lehmann V, Stengl S, Luigart A, (2000) Mat. Sci. Eng. B, 69–70: 11.
[9] Chazalviel J-N, Wehrspohn R, Ozanam F, (2000) Mat. Sci. Eng. B, 69–70: 1.
[10] Langner A, Müller F, Gösele U, (2008) in Molecular- and nano-tubes, eds. Hayden O, Nielsch K, Wang D, .
[11] Canham LT, Parkhutik (Eds) VP, (2000) Phys. Stat. Sol. (a), 182(1).
[12] Canham LT, Nassiopoulou A, Parkhutik V (Eds), (2003) Phys. Stat. Sol. (a), 197.
[13] Canham LT, Nassiopoulou A, Parkhutik V (Eds), (2005) Phys. Stat. Sol. (a), 202(8).
[14] Parkhutik V, Nassiopoulou A, Sailor M, Canham LT (Eds), (2007) Phys. Stat. Sol. (a), 204(5).
[15] Connolly EJ, O'Halloran GM, Pham HTM, Sarro PM, French PJ, (2002) Sens. Actuators A, 99(1–2): 25.
[16] Splinter A, Bartels O, Benecke W, (2001) Sens. Actuators B, 76(1–3): 354.
[17] Hedrich F, Billat S, Lang W, (2000) Sens. Actuators A, 84(3): 315.
[18] Föll H, (1991) Appl. Phys. A, 53: 8.
[19] Propst EK, Kohl PA, (1994) J. Electrochem. Soc., 141(4): 1006.

[20] Ponomarev EA, Levy-Clement C, (1998) Electrochem. Solid-State Lett., 1(1): 42.
[21] Sze SM, (1981) Physics of semiconductor devices, Wiley & Sons, New York.
[22] Chazalviel J-N, (1992) Electrochimica Acta, 37(5): 865.
[23] Lewerenz H-J, (1992) Electrochim. Acta, 37(5): 847.
[24] Langa S, Tiginyanu IM, Carstensen J, Christophersen M, Föll H, (2003) Appl. Phys. Lett., 82(2): 278.
[25] Cojocaru A, Carstensen J, Leisner M, Föll H, Tiginyanu IM, (2008) PSST 2008, accepted.
[26] Cojocaru A, Carstensen J, Föll H, (2008) ECS Trans., 16(3): 157.
[27] Rouquerol J, Avnir D, Fairbridge CW, Everett DH, Haynes JH, Pernicone N, Ramsay JDF, Sing KSW, Unger KK, (1994) Pure & Appl. Chem., 66(8): 1739.
[28] Salem MS, Sailor MJ, Fukami K, Sakka T, Ogata YH, (2008) Phys. Stat. Sol. (a), : to be published.
[29] Christophersen M, Carstensen J, Föll H, (2000) Phys. Stat. Sol. (a), 182(1): 45.
[30] Cojocaru A, Carstensen J, Ossei-Wusu EK, Leisner M, Riemenschneider O, Föll H, (2008) PSST 2008, accepted.
[31] Foca E, Carstensen J, Leisner M, Ossei-Wusu E, Riemenschneider O, Föll H, (2007) ECS Trans., 6(2): 367.
[32] Lehmann V, Föll H, (1990) J. Electrochem. Soc., 137(2): 653.
[33] Fang C, Föll H, Carstensen J, (2006) J. Electroanal. Chem., 589: 259.
[34] Lehmann V, (1993) J. Electrochem. Soc., 140(10): 2836.
[35] Lehmann V, Grüning U, (1997) Thin Solid Films, 297: 13.
[36] Hejjo Al Rifai M, Christophersen M, Ottow S, Carstensen J, Föll H, (2000) J. Electrochem. Soc., 147(2): 627.
[37] Wehrspohn R, Schweizer SL, Geppert T, Lambrecht A, (2008) ECS Trans.
[38] Chazalviel J-N, (1990) Electrochimica Acta, 35(10): 1545.
[39] Carstensen J, Foca E, Keipert S, Föll H, Leisner M, Cojocaru A, (2008) Phys. Stat. Sol. (a), 205(11): 2485.
[40] Popkirov GS, Schindler RN, (1992) Rev. Sci. Instrum., 63: 5366.
[41] Popkirov GS, (1996) Electrochimica Acta, 41(7/8): 1023.
[42] MacDonald JR, (1987) Impedance spectroscopy, John Wiley & Sons.
[43] Leisner M, Carstensen J, Cojocaru A, Föll H, (2008) PSST 2008, accepted.
[44] Leisner M, Carstensen J, Cojocaru A, Föll H, (2008) ECS Trans., 16(3): 133.
[45] Tsuchiya H, Macak J, Taveira L, Schmuki P, (2005) Chem. Phys. Lett., 410: 188.
[46] Lharch M, Aggour M, Chazalviel J-N, Ozanam F, (2002) J. Electrochem. Soc., 149(5): C250.
[47] Langa S, Carstensen J, Tiginyanu IM, Christophersen M, Föll H, (2001) Electrochem. Solid-State Lett., 4(6): G50.
[48] Matthias S, Müller F, Schilling J, Gösele U, (2005) Appl. Phys. A, 80(7): 1391.
[49] Langner A, (2008) Fabrication and characterization of macroporous silicon, Doktorarbeit, Martin-Luther-Universität Halle-Wittenberg.
[50] Müller F, Birner A, Schilling J, Gösele U, Kettner C, Hänggi P, (2000) Phys. Stat. Sol. (a), 182(1): 585.
[51] Christophersen M, Carstensen J, Föll H, (2000) Phys. Stat. Sol. (a), 182(1): 103.
[52] Christophersen M, Carstensen J, Föll H, (2000) Phys. Stat. Sol. (a), 182: 601.
[53] Rönnebeck S, Ottow S, Carstensen J, Föll H, (2000) Journal of Porous Materials, 7: 353.
[54] Chazalviel J-N, Ozanam F, (1992) J. Electrochem. Soc., 139(9): 2501.
[55] Prange R, Carstensen J, Föll H, (1998) in Proc. ECS' 193rd Meeting, San Diego, 158.
[56] Carstensen J, Prange R, Föll H, (1999) J. Electrochem. Soc., 146(3): 1134.
[57] Grzanna J, Jungblut H, Lewerenz HJ, (2000) J. Electroanal. Chem., 486: 181.
[58] Grzanna J, Jungblut H, Lewerenz HJ, (2000) J. Electroanal. Chem., 486: 190.

[59] Grzanna J, Jungblut H, Lewerenz HJ, (2007) Phys. Stat. Sol. (a), 204(5): 1245.
[60] Foca E, Carstensen J, Föll H, (2007) J. Electroanal. Chem., 603: 175.
[61] Carstensen J, Christophersen M, Föll H, (2000) Mat. Sci. Eng. B, 69–70: 23.
[62] Föll H, Carstensen J, Christophersen M, Hasse G, (2000) Phys. Stat. Sol. (a), 182(1): 7.
[63] Lehmann V, Rönnebeck S, (1999) J. Electrochem. Soc., 146(8): 2968.
[64] Lin V S-Y, Motesharei K, Dancil KS, Sailor MJ, Ghadiri MR, (1997) Science, 278: 840.
[65] Janshoff A, Dancil K-P S, Steinem C, Greiner DP, Lin VS-Y, Gurtner C, Motesharei K, Sailor MJ, Ghadiri MR, (1997) J. Am. Chem. Soc., 120(46): 12108.
[66] Vincent G, (1994) Appl. Phys. Lett., 64(18): 2367.
[67] Ishikura N, Fujii M, Nishida K, Hayashi S, Diener J, Mizuhata M, Deki S, (2008) ECS Trans., 16(3): 55.
[68] Propst EK, Rieger MM, Vogt KW, Kohl PA, (1994) Appl. Phys. Lett., 64(15): 1914.
[69] Christophersen M, Carstensen J, Feuerhake A, Föll H, (2000) Mater. Sci. Eng. B, 69–70: 194.
[70] Zheng J, Christophersen M, Bergstrom PL, (2005) Phys. Stat. Sol. (a), 202(8): 1402.
[71] Zheng J, Christophersen M, Bergstrom PL, (2005) Phys. Stat. Sol. (a), 202(8): 1662.
[72] Rieger MM, Kohl PA, (1995) J. Electrochem. Soc., 142(5): 1490.
[73] Ponomarev EA, Lévy-Clément C, (2000) J. Por. Mat., 7: 51.
[74] Wehrspohn RB, Ozanam F, Chazalviel J-N, (1999) J. Electrochem. Soc., 146(9): 3309.
[75] Christophersen M, Carstensen J, Rönnebeck S, Jäger C, Jäger W, Föll H, (2001) J. Electrochem. Soc., 148(6): E267.
[76] Christophersen M, Carstensen J, Voigt K, Föll H, (2003) Phys. Stat. Sol. (a), 197(1): 34.
[77] Wallner JZ, Kunt KS, Obanionwu H, Oborny MC, Bergstrom PL, Zellers ET, (2007) Phys. Stat. Sol. (a), 204(5): 1449.
[78] Frey S, Kemell M, Carstensen J, Langa S, Föll H, (2005) Phys. Stat. Sol. (a), 202(8): 1369.
[79] Takizawa T, Arai S, Nakahara M, (1994) Japan J. Appl. Phys., 33(2, 5A): L643.
[80] Hasegawa H, Sato T, (2005) Electrochim. Acta, 50: 3015.
[81] Tsuchiya H, Hueppe M, Djenizian T, Schmuki P, Fujimoto S, (2004) Sci. Technol. Adv. Mater., 5: 119.

3

Subwavelength Approach of Light Propagation Through Porous Semiconductors

3.1 Introduction

Porous semiconductors with pore sizes below the wavelength of the light offers the opportunity to "engineer" the refractive index at the visible and the IR spectral range by variations of the porosity of the layer. This property can be utilized in a number of optical components [1–3] that will be reviewed later in this book. Note that the luminescence properties of microporous silicon [4] will not be considered here. A lot of research has been dedicated previously to this particular optical use of porous Si and less to other semiconductor materials.

This chapter of the book is devoted to the description of the subwavelength mode of light propagation through porous semiconductors, where the porous semiconductors optically can be considered as effective media. The emphasis is placed on a description of different models, connecting the expected optical properties of porous semiconductors with pore morphology and geometry. Good understanding of the properties of porous semiconductors is essential for two reasons: First, it is necessary for accurate design of mesoporous silicon filters that will be reviewed in Chapter 6. Second, it provides the way for simple and relatively fast identification of some parameters of the morphology of porous semiconductor layers without the need of otherwise inevitable TEM imaging. Review of isotropic effective medium approaches will be given first in this chapter, followed by introduction of anisotropic effective medium models. Last, predictions of some unusual optical properties of certain porous semiconductor materials will be given.

3.2 Isotropic Effective Medium Models

Porous silicon was the first porous semiconductor material whose morphological [5] and optical properties (refractive index and absorption) [6] were studied as early as in 1980s. In a porous medium with the pore dimension or geometry much smaller than the wavelength of light, the dielectric response of the porous medium can be described through an effective dielectric function. In a first order approximation, this effective dielectric function is dependent on the dielectric functions of both the bulk material (semiconductor) and the filling material (e.g. air), as well as on the porosity, while the particular pore morphology is neglected. In the future consideration we will call this approach "isotropic approximation". Such an approximation works reasonably well in the case, mesoporous silicon etched on (100)-oriented Si wafer and near normal incident illumination provides a good example.

Several different effective medium theories were developed in the past. The most popular approaches go back to Bergman [7], Maxwell-Garnett [8], Looyenga [9] and Bruggeman [10]. Effective dielectric permittivities of the porous medium according to the last three approaches are found by solving the following equations:

Bruggeman's EMA: $$f\frac{\varepsilon_I-\varepsilon_{eff}}{\varepsilon_I+2\varepsilon_{eff}}+(1-f)\frac{\varepsilon_B-\varepsilon_{eff}}{\varepsilon_B+2\varepsilon_{eff}}=0 \quad (3.1)$$

Maxwell-Garnett's EMA: $$\varepsilon_{eff}=\varepsilon_B\frac{\varepsilon_I(1+2f)-\varepsilon_B(2f-2)}{\varepsilon_B(2+f)+\varepsilon_I(1-f)} \quad (3.2)$$

Looyenga' EMA: $$\varepsilon_{eff}^{1/3}=(1-f)\varepsilon_B^{1/3}+f\varepsilon_I^{1/3} \quad (3.3)$$

In all the equations ε_{eff} is the effective dielectric permittivity of the composite medium (in our case porous semiconductor), ε_B is the dielectric permittivity of the bulk semiconductor, ε_I is the dielectric permittivity of the inclusion (air in most cases) and f is the porosity.

The main difference between these approaches is in how the microtopology (percolation strength of the network and the sizes of the segments of material left in the medium) is accounted for: In the case of Bruggeman's EMA it is assumed that all the pores (or semiconductor islands) are experiencing an equivalent mean field, which is not the case close to percolation conditions. The assumption of Maxwell-Garnett EMA is the spatial separation of the pores (no intercrossing), which limits its applicability to even lower porosities than Bruggeman's EMA. Neither Maxwell-Garnett nor Looyenga EMA's predict the effective dielectric permittivities at percolation conditions.

Maxwell-Garnett, Looyenga and Bruggeman EMAs offer limited validity for porous semiconductors. This is most clearly seen from the experimentally demonstrated fact that different effective refractive indices can be obtained with porous silicon samples having the same porosity but different pore morphology [11].

Bergman's approach is capable of accounting for the microtopology of the sample by introducing a spectral density function. The problem with this approach is that this function is typically not known *a-priori*. The applicability of different effective medium approaches for porous silicon samples were evaluated by Theiß et al. [12,13]. The results are illustrated in Figure 3.1.

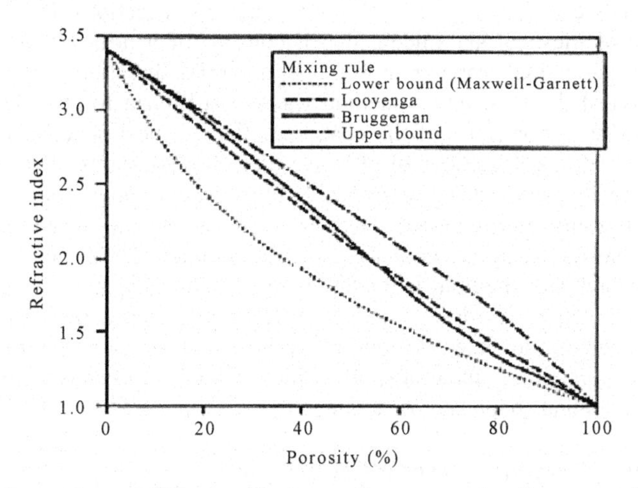

Figure 3.1. Comparison of different effective medium approaches for porous silicon grown on (100)-oriented silicon (after [13])

In general Bruggeman's EMA is the most popular technique when used to predict optical properties of porous semiconductors to date. Validation of its use can be traced back to 1970s [14], when spectroscopic ellipsometry measurements performed on rough amorphous Si were correlated with EMA predictions. However, several authors [15,16] suggest that the Looyenga approach is more accurate for the case of highly porous meso-porous silicon. In general, the difference between the predictions of different theories and the difference between different porous semiconductor samples of the same porosity but different morphology are of the same order of magnitude.

3.3 Anisotropic Effective Medium Models

In the previous paragraph we reviewed the isotropic EMA models. However, recently it has been shown [3,17–20] that mesoporous silicon obtained from a (110) oriented substrate offers optical anisotropy properties that may be of practical importance. A remarkable result is that in the infrared spectral range such a metamaterial offers larger values of optical birefringence than that of any commonly known natural material. Further, the combination of the value of the birefringence in the mid-to far infrared range with the transparency of these materials at those wavelengths makes these findings even more attractive for possible applications and theoretical studies.

Obviously isotropic EMA cannot describe optical anisotropy of certain porous semiconductors. The somewhat generalized Bruggeman EMA [21] was used for calculations of the optical effects in these materials [3,17–19]. This generalization of the EMA, while providing fair estimates for mesoporous (110)-oriented Si, is not applicable for more complex structures, employing e.g. multiple pore lattices, or for materials containing pores with noncircular cross sections. Moreover, even for the case of mesoporous silicon, which consists of a network of pores with circular cross sections growing in some preferential directions, the referenced generalization of the Bruggeman method was not sufficient to provide a full and correct explanation for some of the observed effects. The Looyenga formulation was also applied to calculations of effective dielectric constants of porous silicon [22], however, this method is not applicable to the case of anisotropic medium (at least without some modification). Here we will review our recently suggested method that has the capability to analyze such structures [23]. In such a method the porous semiconductor medium is treated as macroscopically homogeneous and is assigned a dielectric permittivity tensor. Effective dielectric permittivity tensor elements depend on the dielectric constant of the semiconductor host, the dielectric constants of pore-filling materials, the filling fraction (porosity) and the pore parameters including shape and orientation.

3.3.1 Effective Permittivity Tensor Calculations

The model presented here uses several assumptions: First, pores are represented by "air voids"; e.g. by air-filled cavities with elliptical shapes; cf. Figure 3.2. Second, it is assumed that these air voids are uniformly and randomly distributed in the semiconductor material such that the air voids belonging to one of M lattice subsets have their axes essentially parallel to each other. Third, the bulk semiconductor material is assumed to be isotropic and has a dielectric permittivity ε_B. Next, it is assumed that neighboring pores or air holes, respectively, affect each other only through the depolarization factor. This is a strong assumption that limits the validity of the method initially to relatively small porosities (the expanded method of calculation of optical properties of some of porous semiconductors with higher porosities will be presented later in this Chapter). While this latest assumption is not always correct, the accuracy of the approximation presented here is still good enough for a basic understanding of the optical effects. Finally, it is also assumed that the wavelength of the electromagnetic wave considerably exceeds the cross sectional dimensions of the pores.

An electromagnetic wave with the electric field vector $\vec{E}$ gives rise to a displacement vector $\vec{D}$ in the porous semiconductor material, given by

$$\vec{D} = \varepsilon_B \vec{E} + \vec{P} \tag{3.4}$$

where $\vec{P}$ is the effective polarization of all the particles (a pore can always be thought of as a particle in the form of an air void in semiconductor crystal) in a unit volume.

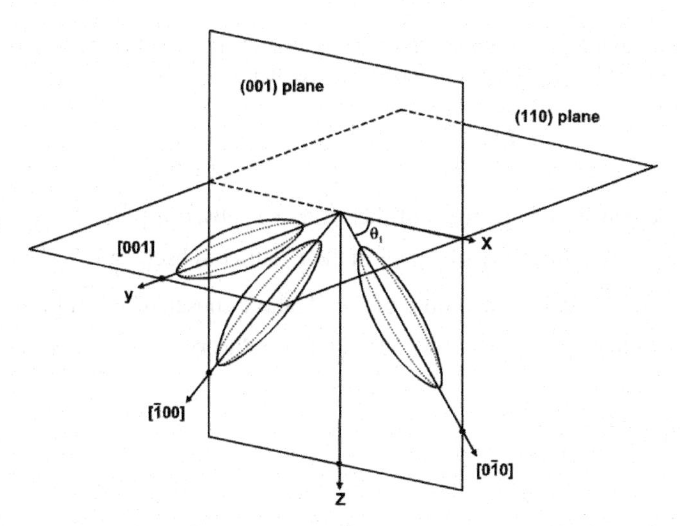

Figure 3.2. Approximating the pore structure of mesoporous silicon grown on (110)-oriented substrate by (ensembles of) ellipsoids. Orientations of three major pore lattice subsets are shown (after [23])

With the assumptions as listed above, the porous semiconductor effective dielectric permittivity tensor $\hat{\varepsilon}^{(eff)} = \begin{pmatrix} \varepsilon_{xx}^{(eff)} & \varepsilon_{xy}^{(eff)} & \varepsilon_{xz}^{(eff)} \\ \varepsilon_{yx}^{(eff)} & \varepsilon_{yy}^{(eff)} & \varepsilon_{yz}^{(eff)} \\ \varepsilon_{zx}^{(eff)} & \varepsilon_{zy}^{(eff)} & \varepsilon_{zz}^{(eff)} \end{pmatrix}$ can be defined as follows:

$$\vec{D} = \hat{\varepsilon}^{(eff)} \vec{E} \tag{3.5}$$

From (3.4) and (3.5) it follows:

$$\hat{\varepsilon}^{(eff)} \vec{E} = \varepsilon_B \vec{E} + \vec{P} \tag{3.6}$$

Within the made assumptions the porous semiconductor material can be considered as an assembly of M electromagnetically separated pore lattice subsets. Hence, the polarization of the porous layer is equal to the vector sum of the polarizations of each lattice subset considered separately, or $\vec{P} = \sum_{i=1}^{M} \vec{P}^{(i)}$, where $\vec{P}^{(i)}$ is the polarization of the ith lattice subset. In this case (3.6) can be rewritten as:

$$\hat{\varepsilon}^{(eff)} \vec{E} = \varepsilon_B \vec{E} + \sum_{i=1}^{M} \vec{P}^{(i)} \tag{3.7}$$

The polarization of each pore is assumed to be a linear isotropic function of the local electric field of the electromagnetic wave, hence:

$$\vec{P}^{(i)} = N^{(i)}\alpha^{(i)}\vec{E}_L^{(i)} \tag{3.8}$$

$N^{(i)}$ is the density of the pores of the ith lattice subset of porous semiconductor layer, $\alpha^{(i)}$ is the polarizability of a pore in the ith lattice subset, and $\vec{E}_L^{(i)}$ is the local electric field as "seen" by each pore of the ith lattice subset. The local field $\vec{E}_L^{(i)}$ is given by Yaghjian [24] for arbitrary shaped inclusions as:

$$\vec{E}_L^{(i)} = \vec{E} + \frac{\hat{L}^{(i)} \bullet \vec{P}^{(i)}}{\varepsilon_B} \tag{3.9}$$

where $\hat{L}^{(i)}$ is the depolarization factor that depends on the shape of the pore; it is a tensor of second rank. Under the assumptions made, the polarizability of each pore is a tensor that is always diagonalizable in a coordinate system, in which one axis coincides with the pore growth direction, i.e., $\hat{\alpha}^{(i)} = \begin{pmatrix} \alpha_{1,1}^{(i)} & 0 & 0 \\ 0 & \alpha_{2,2}^{(i)} & 0 \\ 0 & 0 & \alpha_{3,3}^{(i)} \end{pmatrix}$.

The polarization of the ith lattice subset in the associated coordinate system is $\bar{\vec{P}}^{(i)} = \varepsilon_B \hat{M}^{(i)} \bar{\vec{E}}^{(i)}$, where $\hat{M}_{j,j}^{(i)} = \dfrac{N^{(i)}\alpha_{j,j}^{(i)}}{\varepsilon_B - L_{j,j}N^{(i)}\alpha_{j,j}^{(i)}}$, $j = 1, 2, 3$; and $\hat{M}_{j,k}^{(i)} = 0$, if $j \neq k$.

$\bar{\vec{E}}^{(i)}$ is the electric field of the electromagnetic wave in the coordinate system associated with the ith lattice subset. If the coordinate transformation matrix is introduced between the reference coordinate system (which can be associated with the crystallographic axes of the semiconductor host or anything else) and the coordinate system of ith lattice subset.

$$\hat{A}^{(i)} = \begin{pmatrix} \cos\psi^{(i)}\cos\phi^{(i)} - \cos\vartheta^{(i)}\sin\psi^{(i)}\sin\phi^{(i)} & -\sin\psi^{(i)}\cos\phi^{(i)} - \cos\vartheta^{(i)}\cos\psi^{(i)}\sin\phi^{(i)} & \sin\vartheta^{(i)}\sin\phi^{(i)} \\ \cos\psi^{(i)}\sin\phi^{(i)} + \cos\vartheta^{(i)}\sin\psi^{(i)}\cos\phi^{(i)} & -\sin\psi^{(i)}\sin\phi^{(i)} + \cos\vartheta^{(i)}\cos\psi^{(i)}\cos\phi^{(i)} & -\sin\vartheta^{(i)}\cos\phi^{(i)} \\ \sin\vartheta^{(i)}\sin\psi^{(i)} & \sin\vartheta^{(i)}\cos\psi^{(i)} & \cos\vartheta^{(i)} \end{pmatrix}$$

where φ, ψ and θ are Euler angles, the relations $\bar{\vec{E}}^{(i)} = \hat{A}^{(i)^{-1}}\vec{E}$ and $\vec{P}^{(i)} = \hat{A}^{(i)}\bar{\vec{P}}^{(i)}$ hold where $\vec{E}$ and $\vec{P}^{(i)}$ are in the main coordinate system. Hence

$$\vec{P}^{(i)} = \varepsilon_B \hat{A}^{(i)}\hat{M}^{(i)}\hat{A}^{(i)^{-1}}\vec{E} \tag{3.10}$$

By substituting (3.10) into (3.7) the effective dielectric permittivity tensor of the porous semiconductor layer is obtained as

$$\hat{\varepsilon}^{(eff)} = \varepsilon_B \left[\hat{I} + \sum_{i=1}^{M} \hat{A}^{(i)} \hat{M}^{(i)} \hat{A}^{(i)-1} \right] \tag{3.11}$$

where $\hat{I}_{i,j} = \delta_{i,j}$.

The polarizability tensor $\hat{\alpha}$ and, in principle, the depolarization tensor $\hat{L}$ of individual pores, need to be calculated numerically for the particular pore shape. In [25] the internal field approach combined with Finite Element Method (FEM) has been implemented to find the polarizability tensor elements for circular, square, rectangular and triangular shape of the inclusions and here this approach will be followed.

According to the assumptions listed in the beginning of this section, the pore cross section is assumed to be much smaller than the wavelength of electromagnetic wave (quasi-static approximation). The internal field approach, in which the polarizability is obtained by determining the internal field of the pore, is used so the dipole moment $\vec{d}$ of the air-filled pore in semiconductor material is calculated as

$$\int_V (1-\varepsilon_B)\vec{E}dV \tag{3.12}$$

and the integration is carried out only within the volume of the pore.

From another point of view, the dipole moment $\vec{d}$ is defined as the product of the polarizability α and the local field $\vec{E}_L$. Since it was assumed that the pores are electromagnetically separate, the approximation of a single pore in an infinite medium is accurate enough and, by using the results presented in [26], the dipole moment can be written as:

$$\vec{d} = (1-\varepsilon_B)V \frac{\int_V \vec{E}_{INT} dV}{V} \tag{3.13}$$

where the internal electric field is integrated within the pore. In the dipole approximation, the electric field vector in the pore has x-, y-, z-components, but its integral over the volume of the pore will have only a component for symmetric pores aligned with respect to the electric field. Let's define β as integral of the internal electric field over the pore volume divided by said volume and the external field: $\beta_i = \frac{\int_V \left(\vec{E}_{INT} \cdot \vec{n}_i\right) dv}{V \cdot \left|\vec{E}_{EXT}\right|}$, where $\vec{n}_i$ is a unit vector collinear to i^{th} coordinate axis direction. In the case of electric field alignment along the Cartesian direction j, the dipole moment can be written as:

$$d_j = (1-\varepsilon_B)V\beta_j \vec{E}_{EXT,j} \tag{3.14}$$

The integral β is thus independent on both pore volume and the external electric field of the electromagnetic wave. It can be determined either numerically or analytically. Replacing α_i in (3.8) by $(1-\varepsilon_B)V\beta_i$, the final expression of the dielectric permittivity tensor of porous semiconductor material will be

$$\hat{\varepsilon}^{(eff)} = \varepsilon_B \left[\hat{I} + \sum_{i=1}^{M} \hat{A}^{(i)} \begin{pmatrix} \frac{f_i(1-\varepsilon_B)\beta_1^{(i)}}{\varepsilon_B - L_{11} f_i(1-\varepsilon_B)\beta_1^{(i)}} & 0 & 0 \\ 0 & \frac{f_i(1-\varepsilon_B)\beta_2^{(i)}}{\varepsilon_B - L_{22} f_i(1-\varepsilon_B)\beta_2^{(i)}} & 0 \\ 0 & 0 & \frac{f_i(1-\varepsilon_B)\beta_3^{(i)}}{\varepsilon_B - L_{33} f_i(1-\varepsilon_i)\beta_3^{(i)}} \end{pmatrix} \hat{A}^{(i)-1} \right] \tag{3.15}$$

where f_i is the "porosity" of the ith pore lattice, and $\sum_{i=1}^{M} f_i = p$, $0<p<1$, where p is the total porosity of the porous semiconductor layer.

3.3.2 Effective Permittivity of Mesoporous Silicon Etched on (110)-Oriented Wafers

The method presented above is first applied to analyze the optical effects of mesoporous Si, which can be formed by an electrochemical etching process on (otherwise unusual) (110)-oriented silicon substrates. This porous material has been obtained recently [17] and its optical properties were intensively investigated (see, for example, [3,17–20]). Particularly, strong in-plane birefringence (i.e., optical anisotropy) of such layers has been observed for IR wavelengths. These experimental findings were the basis for several proposed photonic devices based on such a material. For example, dichroic Bragg reflectors have been realized by Diener et al. [3], while optical polarizers have been reported in [19] and [20].

The optical anisotropy of mesoporous silicon fabricated on (110)-oriented substrates has been explained in [17] by the different filling fraction of the porous layer in different directions (i.e., anisotropic porosity). Similar arguments have been presented in [3] to explain the different values of birefringence (i.e., optical anisotropy). In particular, the difference of the porosity in different directions ([001] and $[\bar{1}\bar{1}0]$ crystallographic directions particularly) is proportional to the etching current during mesoporous silicon formation, i.e., proportional to the overall porosity of the sample. The authors of ref. [23] offered a different explanation for the optical properties, better matching the experimental results than the explanation given in [3]. It is based on the formalism reviewed in previous section.

In order to apply the model from ref. [23], the structure of the mesoporous silicon layer grown on (110) substrate have to be identified first. As was shown in [27], pores in mesoporous Si propagate preferentially in equivalent <100>

crystallographic directions independently of the substrate orientation. Hence, it can be represented as a mixture of three lattice subsets of pores collinear to the crystallographic directions mentioned. However, not all <100> directions are equivalent; <100> directions more in line with the electrical field are preferred. This is certainly due to the fact that the electric field strength at the tip then is enlarged, enabling avalanche breakdown [28] and enhancing the electrochemical dissolution reaction at the pore tip.

A representation of the three pore lattice subsets found in mesoporous silicon grown on (110) wafer by some random distribution of many ellipsoids is shown in Figure 3.2. For such a material $[\bar{1}00]$ and $[0\bar{1}0]$ pore directions are equivalent since they have identical projections in the direction of the current flow. The [001] direction, however, is perpendicular to the applied current direction, since it lies in (110) crystallographic plane.

In the analysis presented it is further assumed that the pores in each lattice can be represented as distribution of ellipsoids, which are elongated in the direction of each pore lattice. Such an assumption is in agreement with extensive XRD investigations of different porous layers presented in [29]. In this case Equation 3.11 takes the following form:

$$\hat{\varepsilon}^{(eff)} = \varepsilon_{Si}\left[\hat{I}+\sum_{i=1}^{3}\hat{A}^{(i)}\hat{M}^{(i)}\hat{A}^{(i)-1}\right] \tag{3.16}$$

The pore lattices collinear to the $[0\bar{1}0]$, $[\bar{1}00]$, and [001] crystallographic directions have been assigned index (1), (2), and (3); respectively. It is farther assumed that the individual pores represented by ellipsoids are of the same shape and volume, however, the filling fraction of the pores of the first and second pore lattices exceeds that of the third pore lattice due to the current flow direction, i.e., $f^{(1)} = f^{(2)}$ and $f^{(3)} < f^{(1)}$. With this assumption, depolarization factors and polarizabilities for each pore lattice are the same.

Let's introduce the local pore lattice coordinate system such that the x-axis is parallel to the pore lattice direction and the y-axis lies in the (110) plane. The depolarization factor for the longer axis of the ellipsoid L_{11} depends on the ratio $x = c/a$ $(a > b = c)$ between the axes lengths as (see [36]):

$$L_{11} = \frac{x^2}{\left(1-x^2\right)^{3/2}}\left[arcth\left(\sqrt{1-x^2}\right)-\sqrt{1-x^2}\right] \tag{3.17}$$

Due to the circular cross section of the pores in mesoporous Si, $L_{22} = L_{33} = (1-L_{11})/2$.

If the reference coordinate system is introduced as shown in Figure 3.2 (X-axis is directed in the $[\bar{1}\bar{1}0]$ direction and Y-axis in [001]), the coordinate transformation matrices will be as follows:

$$\hat{A}^{(1)} = \begin{pmatrix} \frac{\sqrt{2}}{2} & 0 & -\frac{\sqrt{2}}{2} \\ 0 & 1 & 0 \\ \frac{\sqrt{2}}{2} & 0 & \frac{\sqrt{2}}{2} \end{pmatrix}, \quad \hat{A}^{(2)} = \begin{pmatrix} -\frac{\sqrt{2}}{2} & 0 & \frac{\sqrt{2}}{2} \\ 0 & -1 & 0 \\ \frac{\sqrt{2}}{2} & 0 & \frac{\sqrt{2}}{2} \end{pmatrix}, \quad \hat{A}^{(3)} = \begin{pmatrix} 0 & 1 & 0 \\ -1 & 0 & 0 \\ 0 & 0 & 1 \end{pmatrix} \tag{3.18}$$

If the porosity of the mesoporous silicon layer is p and the coefficient r will be introduced to describe the ratio of filling fractions between the [001] pore lattice and other lattices, the electrical polarization matrices for each lattice will be as follows:

$$M_{1,1}^{(1)} = M_{1,1}^{(2)} = \frac{(p - p\cdot r)(1-\varepsilon_{Si})\beta_1}{2\varepsilon_{Si} - L_{11}(p - p\cdot r)(1-\varepsilon_{Si})\beta_1}, \quad M_{2,2}^{(1)} = M_{2,2}^{(2)} = M_{3,3}^{(1)} = M_{3,3}^{(2)} = \frac{2(p - p\cdot r)(1-\varepsilon_{Si})\beta_2}{4\varepsilon_{Si} - (1-L_{11})(p - p\cdot r)(1-\varepsilon_{Si})\beta_2}$$

$$M_{1,1}^{(3)} = \frac{p\cdot r\cdot(1-\varepsilon_{Si})\beta_1}{\varepsilon_{Si} - L_{11}\cdot p\cdot r\cdot(1-\varepsilon_{Si})\beta_1}, \quad M_{2,2}^{(3)} = M_{3,3}^{(3)} = \frac{2p\cdot r\cdot(1-\varepsilon_{Si})\beta_2}{2\varepsilon_{Si} - (1-L_{11})\cdot p\cdot r\cdot(1-\varepsilon_{Si})\beta_2}$$

The values of the coefficients β_1 and β_2 according to [36] are:

$$\beta_1 = \frac{\varepsilon_{Si}}{\varepsilon_{Si} - (\varepsilon_{Si} - 1)\cdot L_{11}} \qquad \beta_2 = \frac{\varepsilon_{Si}}{\varepsilon_{Si} - (\varepsilon_{Si} - 1)\cdot L_{22}} = \frac{2\varepsilon_{Si}}{2\varepsilon_{Si} - (\varepsilon_{Si} - 1)\cdot(1 - L_{11})}$$

For the case of an overall porosity of 25%, the aspect ratio of the pore ellipsoids is 0.5 (i.e. longer axis of the ellipsoid is twice longer than other two axes), and the parameter r is 0.1. The refractive index of Si is assumed to be 3.5 with zero imaginary part (which is true for wavelengths exceeding the band edge of Si around 1100 nm and smaller than free carrier absorption edge). Substituting these parameters into (3.15) gives the following value of the dielectric permittivity tensor in the coordinate system as shown in Figure 3.2: $\hat{\varepsilon}^{(eff)} = \begin{pmatrix} 8.421 & 0 & 0 \\ 0 & 8.1 & 0 \\ 0 & 0 & 8.421 \end{pmatrix}$.

This result, based on the effective medium model presented in the previous section, means that mesoporous silicon grown on (110)-substrate will be a negative uniaxial crystal with an optical axis that coincides with the <001> direction. Suffice it to state that the model anisotropy type predicted by the theory presented is in complete agreement with experimental findings as reported in [3] and [17].

The optical anisotropy of such a material is well known (see, for example, [30]). In such crystals for any direction of the electric field in the electromagnetic wave two eigensolutions of the secular equation exist, which are called ordinary and extraordinary waves and described by refractive indices usually denoted n_o and n_e. In the coordinate system as drawn in Figure 3.2, $n_o = \sqrt{\varepsilon_{xx}^{(eff)}} \equiv \sqrt{\varepsilon_{zz}^{(eff)}}$, while $n_e = \sqrt{\varepsilon_{yy}^{(eff)}}$. The normal surface of the electromagnetic waves in this case consists of a sphere and an ellipsoid of revolution, contained in the sphere. For any direction of light propagation in such a crystal two waves for the two different refractive indices exist that have two different polarization states: an ordinary wave, which always related to the refractive index n_o and polarized such that the electric field of the electromagnetic wave is in the (001) plane, and an extraordinary wave related to the refractive index as defined by: $1/n_e^2(\theta) = \cos^2\theta/n_o^2 + \sin^2\theta/n_e^2$, where θ is the angle between the electromagnetic wave propagation direction and the [001] crystal axis. The direction of polarization for the electric field of the extraordinary wave for the case of the plane of incidence being the $(1\bar{1}0)$ plane, is given by $\begin{pmatrix} 0 \\ \cos\theta/(n_e^2(\theta)-n_o^2) \\ \sin\theta/(n_e^2(\theta)-n_o^2) \end{pmatrix}$.

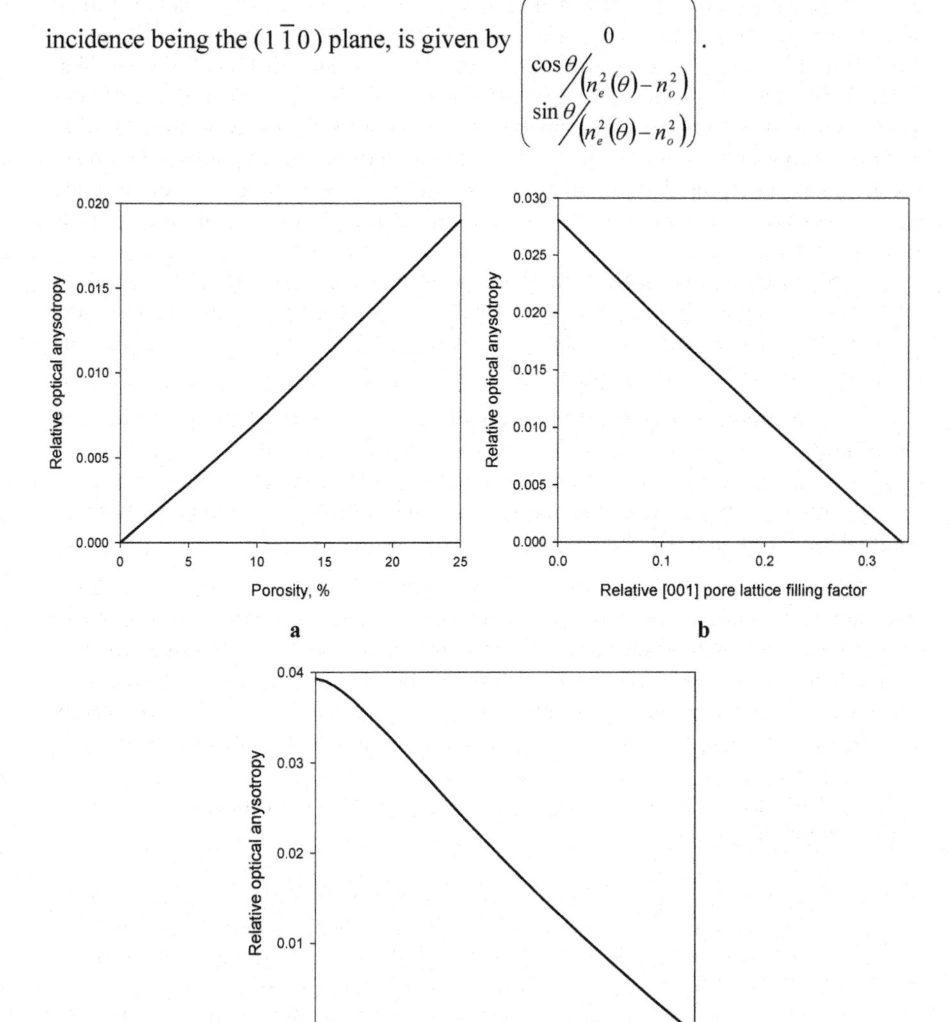

Figure 3.3. Calculated relative optical anisotropy of mesoporous silicon etched on (110) oriented substrate **a** vs. the porosity of the material; **b** vs. the relative filling ratio of the [001] pore lattice of the material, and **c** vs. the pore ellipsoid aspect ratio of the [001] pore lattice of the material (after [23])

The value $(n_o - n_e)/n_o$ of the relative optical anisotropy depends on the porosity of the sample, on the relative filling fraction of the [001] pore lattice subsets, and on the pore ellipsoid aspect ratio. The plots of these dependences are given in Figure 3.3. In all the pictures the porosity of the sample was assumed to be 25%, the aspect ratio of the pore ellipsoids is 0.5, and the relative filling fraction of the [001] pore lattice subset is 0.1 if it is not stated otherwise. The relative optical anisotropy of the material is linearly proportional to the porosity of the sample if both other parameters are assumed to be constant. However, it is highly likely that both parameters can change with experimental conditions that create different porosities in the material. Unfortunately, no extensive experimental investigation of the relative optical anisotropy vs. porosity of the sample is published to date. Such data, interpreted with the formalism presented here, could provide straightforward information on the changes of the morphology of the mesoporous Si layers with porosity.

Based on the results following from the formalism presented in the previous paragraph and as was noted in [23], it is safe to state that optical anisotropy in the mesoporous silicon etched on the (110)-oriented substrate is not due to the anisotropic porosity of the material, but rather due to:

1) anisotropic polarizability and depolarization factors of the pores in each pore lattice,
2) preferential ordering of the pores into three distinct lattices, and
3) smaller filling ratio of the pores aligned into [001] lattice compared to those aligned into [100] and [010] lattices.

The relative optical anisotropy of the material is expected to increase with the porosity of the material even if the relative filling fraction of the [001] pore lattice stays the same. With more experimental data available, the effective medium method theory developed here could help to bring more insights into the pore morphology. The anisotropy of the mesoporous silicon etched on differently oriented substrates can be also easily investigated with the presented methodology.

3.3.3 (100)-Oriented III-V Compound Semiconductors Containing Crystallographic Pores

Other interesting examples of porous materials are porous InP [31–33] and porous GaAs [30,32,34] with crystallographically oriented pores. Optical properties of these materials were theoretically studied by the authors of [35] and we will follow the approach given there. We will consider only the case of porous GaAs, since the case of InP is essentially similar and differs only by the refractive index of the bulk material.

The crystallographically oriented pores in porous GaAs assume the <111>B crystallographic direction as definite growth direction. The zincblende lattice of the III-V compounds consists of double layers with alternating short (three bonds) and long (one bond) distances, and the layers are occupied by A (Ga) or B (As) atoms. The <111>B direction runs from B to A layers along the shortest distance between A and B planes (or from B to A along the longest distance between the A and the B planes). It is important to note that A planes are generally more stable against electrochemical dissolution then B plains. The so-called <111>A directions can be represented as $-Ga \equiv As - Ga \equiv As -$, while the second set (<111>B) can be represented as $-As \equiv Ga - As \equiv Ga -$ (one "-" means one bond).

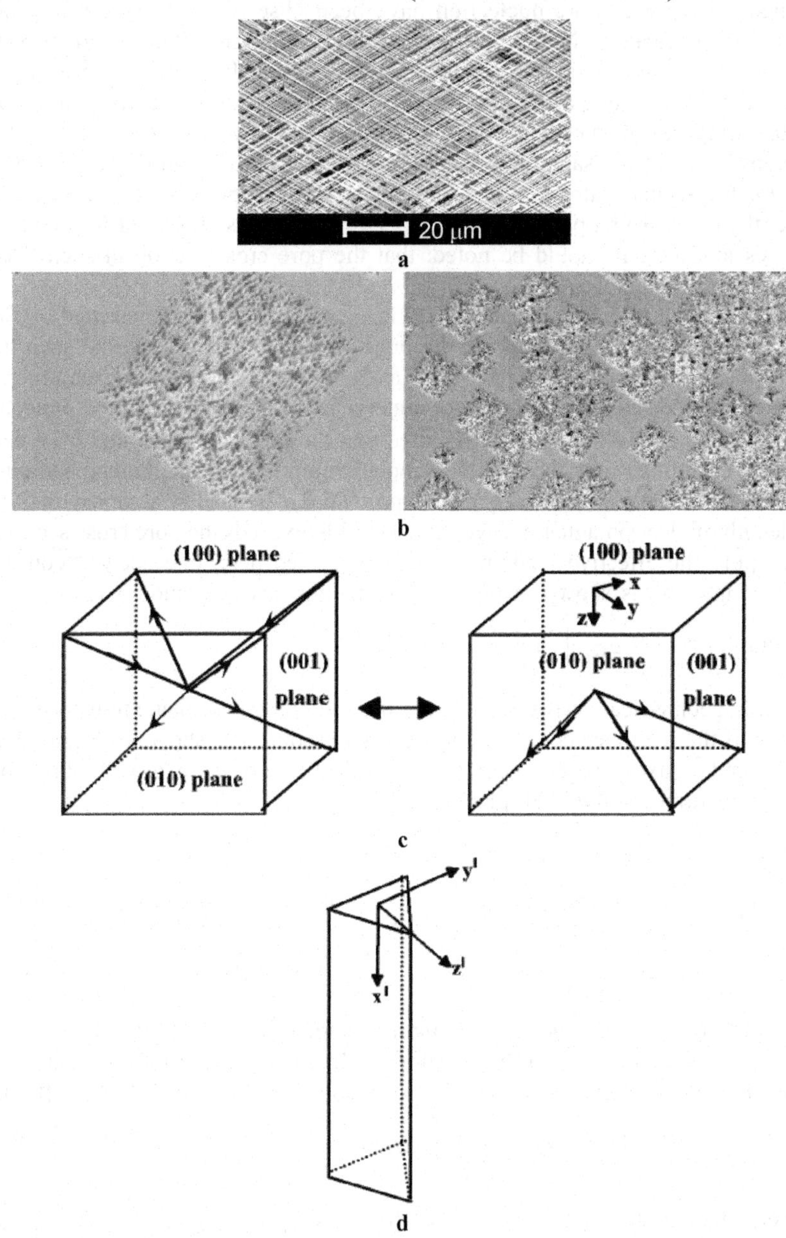

Figure 3.4. a, b SEM images of the porous GaAs layer with crystallographic pores; **c** schematic drawing showing the pore lattices used by the model for the case of electrochemically etched pores in GaAs or InP on (100) oriented substrates, and **d** the coordinate system associated with the pore lattice (after [35])

For an (100)-oriented GaAs wafer, four <111> crystallographic directions are thus available for pore growth. The number of <111>B directions that will be expressed in the pore lattices depends on the nucleation conditions for the pores. Two general types of pore nucleation have been observed [30,36]: uniform and non-uniform nucleation. Non-uniform nucleation usually resulted in formation of "pore domains" on the surface of the sample resulting from pores more or less equally distributed along all <111> directions in the bulk. Uniform nucleation, on the contrary, causing the majority of the pores to grow just in the two <111> directions being in the same plane and going "downwards", while the density of pores for the second pair of <111> directions (going "upwards") is considerably smaller than that of first pair. Figures 3.4a–3.4c show cross section SEM images of the pores in GaAs. It should be noted, that the pore cross-section in such GaAs layers has a triangular shape.

In the model it is assumed that porous GaAs material can be represented by four sublattices of elongated air voids in the bulk of semiconductor material such that the air voids in each sublattice have their axes essentially parallel to each other. It is further assumed that the bulk GaAs material is isotropic and has the dielectric permittivity ε_{GaAs}. Further on, it is assumed that the pores are separated from each and affect each other only through the depolarization factor (this limits the validity of the used method to relatively small porosities). It is also assumed that the wavelength of electromagnetic waves considerably exceeds the pore cross section.

To apply the effective medium method presented here, one need to consider four pore lattices, as shown in Figure 3.4c. In this case Equation 3.11 takes the following form: $\hat{\varepsilon}^{(eff)} = \varepsilon_{GaAs}\left[\hat{I} + \sum_{i=1}^{4} \hat{A}^{(i)} \hat{M}^{(i)} \hat{A}^{(i)-1}\right]$.

Simple geometrical derivations lead to the following coordinate transformation matrices, if the reference coordinate system is introduced as shown in Figure 3.4c. The X- and Y-axes are collinear to the projections of pore growth directions (<111> directions) on the (100) plane:

$$\hat{A}^{(1)} = \begin{pmatrix} \sqrt{\frac{2}{3}} & 0 & -\frac{1}{\sqrt{3}} \\ 0 & 1 & 0 \\ \frac{1}{\sqrt{3}} & 0 & \sqrt{\frac{2}{3}} \end{pmatrix}, \hat{A}^{(2)} = \begin{pmatrix} 0 & 1 & 0 \\ -\sqrt{\frac{2}{3}} & 0 & \frac{1}{\sqrt{3}} \\ \frac{1}{\sqrt{3}} & 0 & \sqrt{\frac{2}{3}} \end{pmatrix}, \hat{A}^{(3)} = \begin{pmatrix} -\sqrt{\frac{2}{3}} & 0 & \frac{1}{\sqrt{3}} \\ 0 & -1 & 0 \\ \frac{1}{\sqrt{3}} & 0 & \sqrt{\frac{2}{3}} \end{pmatrix}, \hat{A}^{(4)} = \begin{pmatrix} 0 & -1 & 0 \\ \sqrt{\frac{2}{3}} & 0 & -\frac{1}{\sqrt{3}} \\ \frac{1}{\sqrt{3}} & 0 & \sqrt{\frac{2}{3}} \end{pmatrix}$$

In order to understand how the triangular shape of the pores affects the optical anisotropy of the porous GaAs material, first the dielectric permittivity tensor for pores with a circular cross section will be calculated. In this case [37], $\beta_{22} = \beta_{33}$ and $L_{22} = L_{33} = (1-L_{11})/2$; $L_{11} = \frac{x^2}{(1-x^2)^{3/2}}\left[arcth\left(\sqrt{1-x^2}\right) - \sqrt{1-x^2}\right]$, with x = the ratio between the axes length; $x = c/a$ $(a > b = c)$; $\beta_{11} = \frac{\varepsilon_{Si}}{\varepsilon_{Si} - (\varepsilon_{Si} - 1)\cdot L_{11}}$; $\beta_{22} = \beta_{33} = \frac{2\varepsilon_{Si}}{2\varepsilon_{Si} - (\varepsilon_{Si} - 1)\cdot(1 - L_{11})}$. As follows from the SEM images (see Figure

3.4a), the pores in GaAs are best represented as cylinders, or as the limiting case of ellipsoids with $x \rightarrow 0$.

The results of dielectric permittivity tensor elements calculations are presented in Figure 3.5 (dashed lines). The electromagnetic wave was assumed to propagate along the <100> direction with the electric field vector of the electromagnetic wave coinciding with the <011> crystallographic direction. A 30% overall porosity of the GaAs layer is assumed. Calculations show that in this case, a porous GaAs layer would exhibit a biaxial anisotropy for all relative filling fractions of two pore lattice pairs except for the case of equal filling, where it would behave as an optically isotropic material.

Now we take into account the triangular shape of the pore cross-section. For this one needs to reevaluate the values of the L_{ii} and β_i coefficients. Since the ratio of pore length to pore cross-section for porous GaAs is very high, with good accuracy, $L_{11} = 0$ and $\beta_{11} = 1$. For equilateral triangular shape inclusions [36] $L_{22} = 0.5 \Rightarrow L_{33} = 1 - L_{22} = 0.5$. β_2 was assumed to be 1.82, while $\beta_3 = 1.86$.

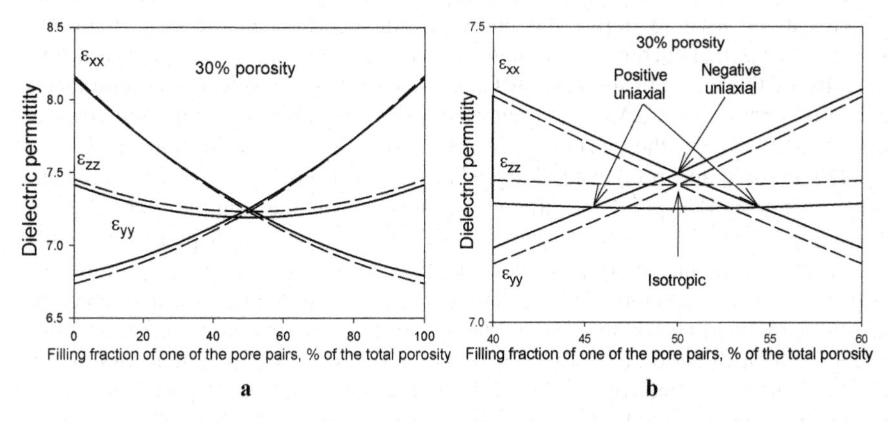

Figure 3.5. Calculated dielectric tensor elements of the porous GaAs layer with crystallographic pores etched on (100)-oriented substrate as a function of the relative filling fraction of one of the pore lattice pairs: **a** general view; **b** magnified view near the 50% filling fraction. The overall porosity of the sample is 30%. Dashed lines indicate the results of calculations for the case of the pores with circular cross section, solid lines indicate the results of calculations with the account of the triangular cross sections of crystallographic pores in GaAs (after [34])

These corrections account for the triangular shape of the pore cross sections and allow calculations of the dielectric permittivity tensor elements of porous GaAs layer with the same set of parameters that was used for the calculations for the circular pore cross section. The results of such calculations are also presented in Figure 3.5. The triangular shape of the pore causes quite interesting effects, which can be viewed better in Figure 3.5b. For example, for pore lattice pairs with identical representation or "weight", i.e. a relative filling fraction of 50%, porous GaAs no longer exhibits isotropic behavior, but rather becomes a negative uniaxial crystal with an optical axis that coincides with the <100> crystallographic direction. Besides the case of both pore lattice pairs being equally represented, two more points of non-biaxial behavior appear. For the parameters of the porous GaAs layer as used for the calculations, these points correspond to filling fractions of roughly 45.5% and 54.5%. At these points the porous GaAs layer will exhibit positive uniaxial behavior with optical axes aligned along the two perpendicular [011] crystallographic directions, respectively. In all other cases the porous GaAs layer will exhibit biaxial behavior. Hence, as was noted in [34], porous GaAs (or InP and so on) provides the unique capability to control not only the optical anisotropy value, as, for example, in mesoporous silicon etched on (110) oriented substrates [3,17–20], but also the optical anisotropy type (uniaxial or biaxial), and even the direction of the optical axis by modifying the etching parameters.

The challenge is to develop the right porosity, cross-section, and filling factors of the pore lattices of the porous GaAs for the desired optical properties. The morphology of a porous semiconductor depends on the electrochemical etching conditions used. Beside parameters like electrolyte composition, applied current density or etching bias, the nucleation process plays a critical role for the overall morphology. The overall current density defines the amount of dissolved material, which can be used to define the porosity. For example, in the case of porous silicon, the applied current determines under certain etching conditions directly the porosity of the layer. In the case of GaAs this approach is not as straightforward because pores in GaAs can nucleate in the form of pore-domains or homogeneously on the surface. As mentioned, the pores in both cases are crystallographic and propagate along <111>B directions. In both cases from one primary nucleation point on the (100) surface of the sample two primary crystallographic pores start to grow. If the primary nucleation points are dense enough, no branching of the two nucleated pores is observed, and a uniform 3D structure will develop. The "cross-hatched" structure occurs because crystallographically oriented pores are able to intersect each other without changing their direction of growth [35]. On the other hand, if the nucleation points are less dense, multiple branching of the two initially nucleated pores and of the secondary pores created by branching will occur; the result is a pore domain with a particular structure. For a homogeneous nucleation it is necessary to obtain a high density of nucleation points on the surface of the sample, which can be formed by a two-step anodization process [38]. Filling factors other then 50 % might be obtained by starting with a substrate that is slightly off the (100) orientation, because the more steeply inclined growth directions are often favored in pore growth. Further details on the different anodization conditions of porous GaAs can be found elsewhere [30,35,37,39,40].

3.3.4 Surface Plasmon Enhancement of an Optical Anisotropy in Porous Silicon/Metal Composites

Metal filling of the pore arrays through, e.g., electroplating or electroless plating techniques, has been already demonstrated [41,42]. Some of the metals, e.g. silver, gold, aluminum, copper, are known as a good providers of surface plasmons (SPs), with non-radiative waves existing on the metal-semiconductor interfaces. SPs are known to be responsible for large enhancements of the electromagnetic field and to influence or dominate the scattering cross-section of metal colloids [43], Raman scattering, higher harmonic generation [44], etc. The question coming up now is how the linear optical properties (e.g., anisotropy) of the porous semiconductor (such as for example mesoporous silicon etched on (110) substrate) will change if the pores will be filled with SP-active metals. This question was studied theoretically in [44]. The approach, developed in [44] is also useful from a methodological standpoint, since it introduces a modification of Bruggeman's EMA that is expected to provide better accuracy for the prediction of optical properties of porous semiconductors with unfilled pores as well.

The structure of the mesoporous silicon layer grown on a (110) substrate was discussed previously in this Chapter. First, let's apply the theory presented earlier in this chapter developed for multiple sub-lattices of weakly polarizable inclusions. Equation 3.11 then will take the form:

$$\hat{\varepsilon}^{(eff)} = \varepsilon_B \left[\hat{I} + \sum_{i=1}^{3} \hat{A}^{(i)} \hat{M}^{(i)} \hat{A}^{(i)^{-1}} \right] \tag{3.19}$$

where ε_m is the dielectric permittivity of the metal filling the pores. If the porosity of the mesoporous silicon layer is p and the coefficient r will be introduced to describe the ratio of filling fractions between the [001] metal-filled pore lattice and other two lattices, the electrical polarization matrices $M^{(i)}$ for each lattice will be as follows:

$$M^{(1)} = M^{(2)} = \begin{pmatrix} \frac{(p-p\cdot r)\alpha_{1,1}^{(i)}}{2\varepsilon_{Si} - L_{1,1}(p-p\cdot r)\alpha_{1,1}^{(i)}} & 0 & 0 \\ 0 & \frac{2(p-p\cdot r)\alpha_{2,2}^{(i)}}{4\varepsilon_{Si} - (1-L_{1,1})(p-p\cdot r)\alpha_{2,2}^{(i)}} & 0 \\ 0 & 0 & \frac{2(p-p\cdot r)\alpha_{3,3}^{(i)}}{4\varepsilon_{Si} - (1-L_{1,1})(p-p\cdot r)\alpha_{3,3}^{(i)}} \end{pmatrix}$$

$$M^{(3)} = \begin{pmatrix} \frac{p\cdot r\cdot \alpha_{1,1}^{(3)}}{\varepsilon_{Si} - L_{1,1}\cdot p\cdot r\cdot \alpha_{1,1}^{(3)}} & 0 & 0 \\ 0 & \frac{2p\cdot r\cdot \alpha_{2,2}^{(3)}}{2\varepsilon_{Si} - (1-L_{1,1})\cdot p\cdot r\cdot \alpha_{2,2}^{(3)}} & 0 \\ 0 & 0 & \frac{2p\cdot r\cdot \alpha_{3,3}^{(3)}}{2\varepsilon_{Si} - (1-L_{1,1})\cdot p\cdot r\cdot \alpha_{3,3}^{(3)}} \end{pmatrix} \tag{3.20}$$

If the metal inclusions in different subsets have the same shape and volume but different orientations, as was assumed originally, we have $\alpha_{i,j}^{(k)} = \alpha_{i,j}^{(l)}$.

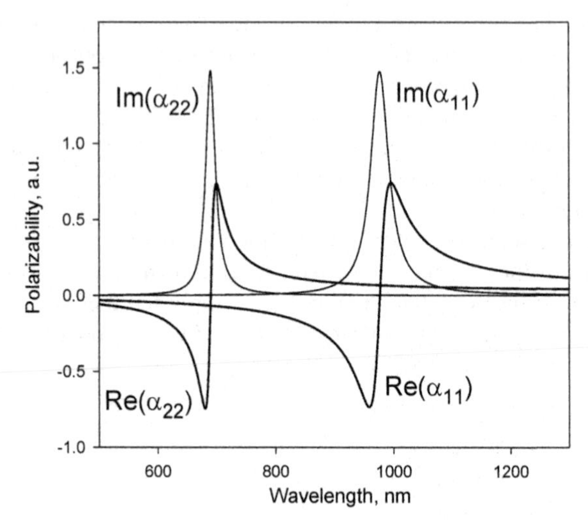

Figure 3.6. Calculated spectral dependences of the polarizability tensor coefficients of the elongated silver ellipsoid with a = 15 nm and b = 10 nm, embedded into the silicon host. After [45])

To determine theses coefficients $\alpha_{i,j}^{(k)}$ we will use the result of Kuwata [42] for calculations of the normalized (by volume) polarizabilities of metal ellipsoids, which states

$$\alpha_{i,i}(\varepsilon) = \frac{1}{(L_{i,i} + \frac{\varepsilon}{\varepsilon_m - \varepsilon}) + A_{i,i}\varepsilon_m x_i^2 + B_{i,i}\varepsilon_m^2 x_i^4 - i\frac{4}{3}\cdot\pi^2\varepsilon_m^{2/3}\frac{V}{\lambda^3}} \tag{3.21}$$

where $x_1 = \pi a/\lambda$; $x_2 = x_3 = \pi b/\lambda$; $x = a/b$;

$A_{i,,i} = -0.4865L_{i,i} - 1.046L_{i,i}^2 + 0.8481L_{i,i}^3$,

$B_{i,,i} = 0.01909L_{i,i} + 0.1999L_{i,i}^2 + 0.6077L_{i,i}^3$; λ is the wavelength.

Equation 3.21, according to [42], is valid for metal inclusion dimensions with sizes in the range from just a few nanometers to at least some hundreds of nanometers.

The spectral dependences of the polarizability coefficients for the case of a silver elongated ellipsoid with a = 15 nm and b = 10 nm, embedded into a silicon host, calculated with Equation 3.21 are given in Figure 3.6.

The dielectric constant of silver was calculated according to the Drude approximation. For the calculations the imaginary part of the silicon refractive index at the given wavelengths was neglected. The reason for this is that in porous silicon the position of the absorption edge is strongly blue-shifted (see [46]), so such an assumption indeed makes sense. As one can expect, the polarizability resonances of the silver ellipsoids for electric fields of the electromagnetic wave aligned along different ellipsoid axes are located at different wavelengths. These polarizability resonances are related to the excitation of the surface plasmon modes on the ellipsoid surface.

By substituting (3.21) into (3.20) and calculating the coefficients of the dielectric permittivity tensor of silver-porous silicon etched on (110) substrate composite, one can show that the resultant material will exhibit an uniaxial type of anisotropy with the optical axis coinciding with the <001> silicon crystallographic direction, the same as for unfilled pores considered previously in this Chapter.

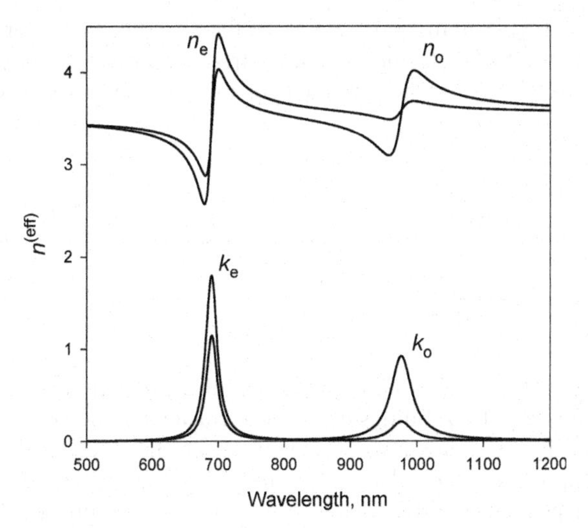

Figure 3.7. Calculated spectral dependences of the real (n) and imaginary (k) parts of ordinary and extraordinary refractive indices of the silver/porous silicon etched on (110)-oriented wafer with filling fraction of the silver ellipsoids of 0.1 volume percent (after [44])

The calculated spectral dependences of the real (n) and imaginary (k) parts of ordinary and extraordinary refractive indices of the material with filling fraction of the silver ellipsoids of just 0.1 volume percent are given in Figure 3.7. One can see that the model predicts a substantial optical anisotropy of such a material, even for the small filling fractions used, and that the uniaxial anisotropy of the material changes the "sign" (from positive to negative) around the positions of the surface plasmon resonances.

It should be noted that the model presented above can provide reasonable results only for very small filling fractions of metal ellipsoids, especially around the resonances. Moreover, such a model can only predict the dielectric properties of composites consisting of weakly-polarizable inclusions fairly realistically; for such highly-polarizable inclusions as metal ellipsoids around plasmon resonance excitation conditions, the model is not really adequate. More self-consistent models need to be implemented and the following material is devoted to this task.

The authors of [44] suggested a generalization of the Bruggeman [21] method for the case of composites consisting of multiple sublattices of inclusions. The main assumption in the Bruggeman method is that the total electrical polarizability of the composite material is equal to zero. In other words, the total polarizability of all the inclusions (in the analyzed case both metal inclusions and silicon "inclusions") embedded in the medium with effective dielectric parameters should vanish:

$$\sum_{i=1}^{3} \tilde{\vec{P}}^{(i)} + \tilde{\vec{P}}^{(4)} = 0 \qquad (3.22)$$

where $\tilde{\vec{P}}^{(i)}$, i = 1, 2, 3 is the polarization of the ith lattice subset and $\tilde{\vec{P}}^{(4)}$ is the polarizability of the Si "inclusion" in the matrix. According to the abbreviations introduced above, (3.22) can be rewritten for the analyzed case in the following form:

$$(1-p)\hat{\alpha}^{(4)}\vec{E} + \frac{p-p\cdot r}{2}\hat{A}^{(1)}\hat{\alpha}^{(1)}\hat{A}^{(1)^{-1}}\vec{E} + \frac{p-p\cdot r}{2}\hat{A}^{(2)}\hat{\alpha}^{(2)}\hat{A}^{(2)^{-1}}\vec{E} +$$
$$+ p\cdot r\cdot \hat{A}^{(3)}\hat{\alpha}^{(3)}\hat{A}^{(3)^{-1}}\vec{E} = 0 \qquad (3.23)$$

Generally, Equation 2.23 states a fairly complex problem since it includes the determination of the polarizabilities of the all inclusions in the anisotropic medium with generally speaking unknown orientations of the axes, and thus requires solving a system of six independent equations with six unknowns (according to the number of independent coefficients in the dielectric permittivity tensor of the effective medium). However, in the analyzed case one can simplify the solution of problem (3.23) by using the results obtained from the generalization of the Maxwell-Garnett model already introduced in section.

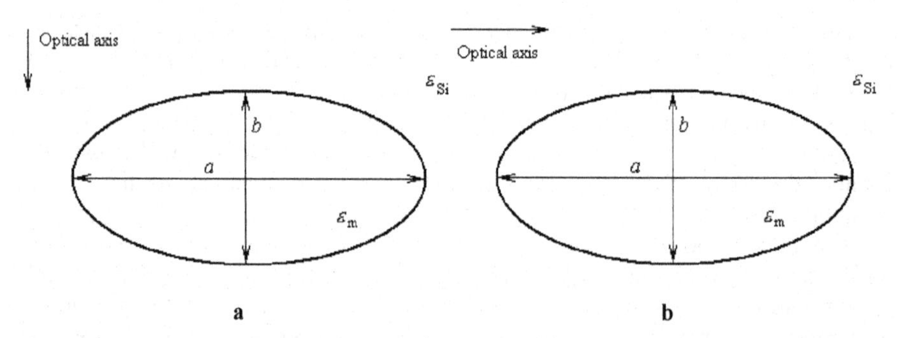

Figure 3.8. Orientation of the optical axis of the metal-filled porous silicon etched on (110) oriented substrate with respect to the major axis of the meal inclusion ellipsoid **a** for 1st and 2nd inclusion subsets, and **b** for 3rd inclusion subsets (after [44])

Although, as mentioned before, this model cannot provide sufficiently accurate quantitative results for the material analyzed here, the qualitative predictions of an uniaxial type of anisotropy with optical axis aligned with the <001> crystallographic direction can certainly be trusted. Taking this into account, the polarizability tensors of metal inclusions in each lattice subset can be written as:

$$\hat{\alpha}^{(1)} = \begin{pmatrix} \alpha_{1,1}(\varepsilon_o) & 0 & 0 \\ 0 & \alpha_{2,2}(\varepsilon_e) & 0 \\ 0 & 0 & \alpha_{3,3}(\varepsilon_o) \end{pmatrix} \hat{\alpha}^{(2)} = \begin{pmatrix} \alpha_{1,1}(\varepsilon_o) & 0 & 0 \\ 0 & \alpha_{2,2}(\varepsilon_e) & 0 \\ 0 & 0 & \alpha_{3,3}(\varepsilon_o) \end{pmatrix}$$

$$\hat{\alpha}^{(3)} = \begin{pmatrix} \alpha_{1,1}(\varepsilon_e) & 0 & 0 \\ 0 & \alpha_{2,2}(\varepsilon_o) & 0 \\ 0 & 0 & \alpha_{3,3}(\varepsilon_o) \end{pmatrix} \tag{3.24}$$

where the $\alpha_{i,j}^{(k)}$ coefficients are determined according to (3.21). This can be done because the optical axis of the effective medium is always aligned with one of the axes of an inclusion ellipsoid for porous silicon etched on (110)-oriented substrate, as illustrated in Figure 3.8a for inclusions from 1st and 2nd subsets and Figure 3.8b for inclusions from 3rd subset. Silicon "particles" in this approximation can be assumed to have a spherical shape. It makes sense to preserve the reference coordinate system in the silicon "particles" matrix. In this case,

$$\hat{\alpha}^{(4)} = \begin{pmatrix} \dfrac{\varepsilon_{Si} - \varepsilon_o}{\varepsilon_{Si} + 2\varepsilon_o} & 0 & 0 \\ 0 & \dfrac{\varepsilon_{Si} - \varepsilon_e}{\varepsilon_{Si} + 2\varepsilon_e} & 0 \\ 0 & 0 & \dfrac{\varepsilon_{Si} - \varepsilon_o}{\varepsilon_{Si} + 2\varepsilon_o} \end{pmatrix}.$$

Hence, Equation 3.23 can be rewritten into a system of two equations with two unknowns (ε_o and ε_e):

$$\begin{cases} (1-p)\alpha_{1,1}^{(4)} + \dfrac{p - p\cdot r}{4}\left(\alpha_{1,1}^{(1)} + \alpha_{3,3}^{(1)} + \alpha_{1,1}^{(2)} + \alpha_{3,3}^{(2)}\right) + p\cdot r\cdot \alpha_{2,2}^{(3)} = 0 \\ (1-p)\alpha_{2,2}^{(4)} + \dfrac{p - p\cdot r}{2}\left(\alpha_{2,2}^{(1)} + \alpha_{2,2}^{(2)}\right) + p\cdot r\cdot \alpha_{1,1}^{(3)} = 0 \end{cases} \tag{3.25}$$

According to the assumptions the metal "particles" in each of the lattices have the same shape and volume. Equation 3.25 can then be further simplified to:

$$\begin{cases} (1-p)\alpha_{1,1}^{(4)} + \dfrac{p - p\cdot r}{2}\left(\alpha_{1,1}^{(1)} + \alpha_{3,3}^{(1)}\right) + p\cdot r\cdot \alpha_{2,2}^{(3)} = 0 \\ (1-p)\alpha_{2,2}^{(4)} + (p - p\cdot r)\alpha_{2,2}^{(1)} + p\cdot r\cdot \alpha_{1,1}^{(3)} = 0 \end{cases} \tag{3.26}$$

Equation 3.26 can be easily solved numerically.

Based on this model, Figure 3.9 gives the calculated spectral dependences of the refractive indices of the ordinary and extraordinary waves for the silver/porous silicon system on (110)-oriented substrate composites. Figure 3.9a and 3.9c give the spectral dependences of the real parts of the refractive indices for the composites with 1% and 0.1% metal filling fractions respectively, while Figure 3.9b and 3.9d give the spectral dependences of the imaginary parts of the refractive indices (also known as attenuation coefficients). In all the figures the silver ellipsoids were assumed to be of an elongated shape with a longer axis of 15 nm and shorter axes of 10 nm.

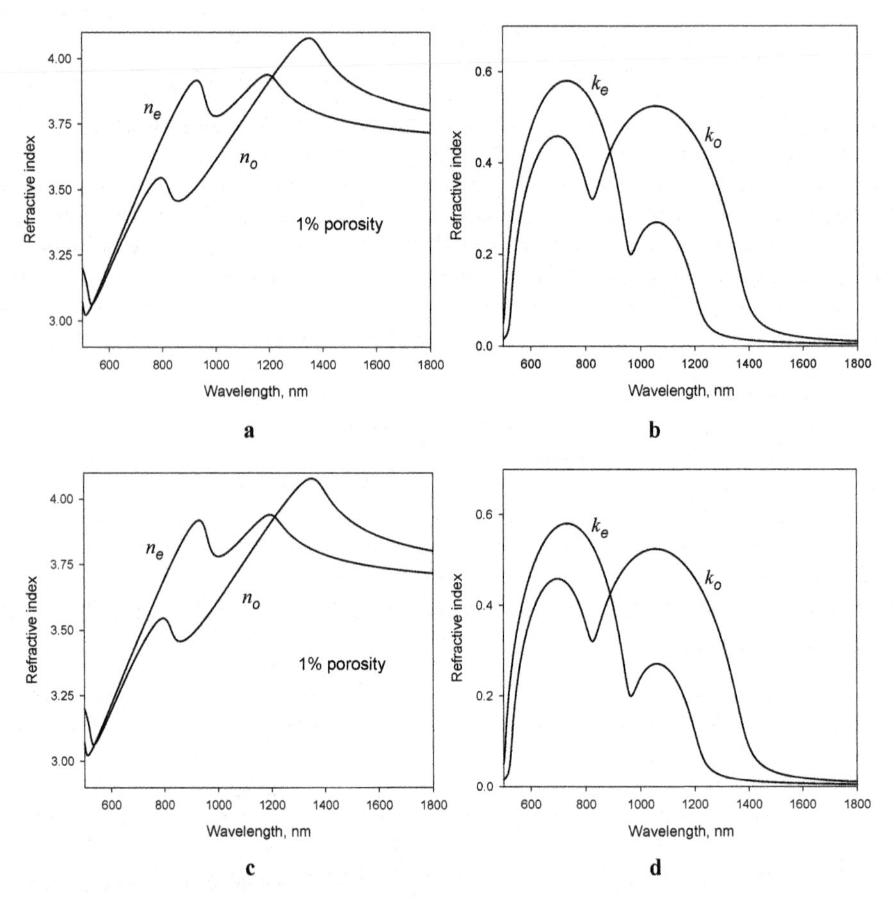

Figure 3.9. Calculated spectral dependences of the **a** real and **b** imaginary parts of the refractive indices of ordinary and extraordinary waves of the silver-filled porous silicon etched on (110)-oriented substrate for the case of 1% filling fraction of metal; and numerically calculated spectral dependences of the **c** real and **d** imaginary parts of the refractive indices of ordinary and extraordinary waves of the silver-filled porous silicon etched on (110)-oriented substrate for the case of 0.1% filling fraction of metal (after [44])

The parameter r (see above) was assumed to be 0.1. One can see that the optical anisotropy of such a material is indeed expected to be quite high. While at 0.1% filling fraction the calculated dependences (Figures 3.9c and 3.9d) of the reflective indices closely resemble those obtained with the generalized Maxwell-Garnett methodology (Figure 3.7), for filling fractions still as small as 1%, the spectral dependences are changing shape dramatically. However, the surface plasmon enhancement of the optical anisotropy at certain wavelengths is still very strong.

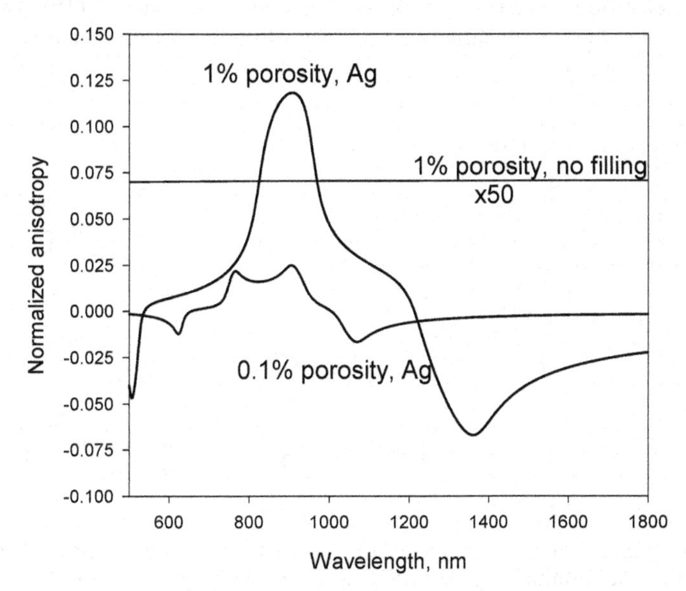

Figure 3.10. Calculated spectral dependences of the relative optical anisotropy of the silver-filled porous silicon etched on (110)-oriented substrate for 0.1% and 1% filling fraction of silver, and of the air-filled porous silicon etched on (110)-oriented substrate with 1% porosity (multiplied by 50). After [44]

Figure 3.10 presents the spectral dependences of the relative optical anisotropies ($|n_o - n_e|/n_o$) of the silver-filled porous silicon etched on (110)-oriented substrate for 0.1% and 1% filling fraction of silver and of the air-filled porous silicon etched on (110)-oriented substrate with 1% porosity. An almost 100-fold enhancement of the optical anisotropy is predicted due to the surface plasmon resonance on silver ellipsoids.

The analyzed material can also exhibit quite interesting properties at intermediate filling fractions. For example, Figure 3.11 shows the numerically calculated spectral dependences of the dielectric permittivity tensor coefficients for the silver-filled porous silicon etched on (110)-oriented substrate with 17% volume fraction of metal. Figure 3.11a gives the spectral dependences of the real parts of the effective dielectric permittivity tensor coefficients, while Figure 3.11b presents the spectral dependences of the imaginary parts of those. The model predicts different signs of diagonal elements of effective dielectric permittivity tensor around 1250–1775 nm wavelengths. Materials having these parameters within this spectral range belong to the class of materials with "indefinite" permittivity tensors, using the terminology introduced in [47]. It turns out that "indefinite" materials based on porosity offer quite peculiar optical properties. For example, some of the optical modes (surface or bulk) in such materials exist only for certain directions of propagation, but disappear in other directions, a property that can be used in e.g. spatial filtering. However, it should be noted that according to the model presented here, the losses in the analyzed material within the "indefinite" parameter range will be high (see the large anisotropic imaginary part of the effective dielectric permittivity tensor elements given in Figure 3.11b). Though, it should be noted that the quantitative accuracy of the model presented deteriorates at such concentrations.

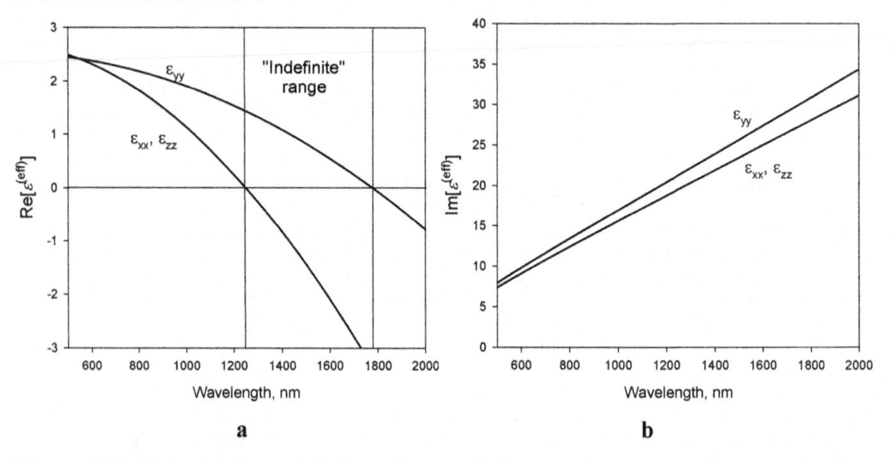

Figure 3.11. Calculated spectral dependences of the **a** real and **b** imaginary parts of the effective dielectric permittivity tensor coefficients of the silver-filled porous silicon etched on (110)-oriented substrate for the case of a 17% filling fraction of the metal (after [44])

Another interesting effect that should take place in the metal-filled porous silicon etched on (110)-oriented substrates is an enhancement of various nonlinear effects, such as Raman scattering and higher order harmonic generation (second, third harmonics, etc.). Such effects should be greatly enhanced due to very high enhancement of the electromagnetic field in the vicinity of metal inclusions. Moreover, uniaxial anisotropy of the analyzed material should make it possible to achieve phase matching (a must in order to get strong coupling between irradiated and double frequency waves).

3.4 Conclusions

Subwavelength regime of light propagation through porous semiconductors offers a wide range of interesting phenomena. As was reviewed in this chapter, the porous semiconductors not only offer possibilities to the engineer the isotropic optical properties of semiconductor materials (e.g., refractive index and absorption coefficients), but also permit to engineer the optical anisotropy as well. Uniaxial anisotropy as well as biaxial anisotropy are predicted for different porous semiconductors. Moreover, porous semiconductors can serve as templates for even more unusual materials (e.g., metal-filled mesoporous silicon etched on (110)-oriented substrates), so-called indefinite materials. While it has already been demonstrated that isotropic optical properties of porous silicon have an application potential in optical filters (that will be reviewed in detail in Chapter 6), the potential applications of anisotropic porous semiconductors are not that well defined as of yet. Polarizers, wave-plates, imaging elements may be just a small subset of what can be made with these materials.

3.5 References

[1] Vincent G, (1994) Optical properties of porous silicon superlattices. Appl. Phys. Lett. 64:2367–2369.
[2] Pellegrini V, Tredicucci A, Mazzoleni C, Pavesi L, (1995) Enhanced optical properties in porous silicon microcavities. Phys. Rev. B 52:R14328.
[3] Diener J, Künzner N, Kovalev D, Gross E, Timoshenko VY, Polisski G, Koch F, (2001) Dichroic Bragg reflectors based on birefringent porous silicon. Appl. Phys. Lett. 78:3887–3889.
[4] Canham LT, (1990) Silicon quantum wire array fabrication by electrochemical and chemical dissolution of wafers. Appl. Phys. Lett. 57:1046–1048.
[5] Beale MIJ, Benjamin JD, Uren MJ, Chew NG, Cullis AG, (1985) An experimental and theoretical-study of the formation and microstructure of porous silicon. J. Cryst. Growth, 73:622–636.
[6] Pickering C, Beale MIJ, Robbins DJ, Pearson PJ, Greef R, (1985) Optical-properties of porous silicon films. Thin Solid Films, 125:157–163.
[7] Bergman DJ, (1978) Dielectric-constant of a composite-material – problem in classical physics. Phys. Rep., 43:378–407.
[8] Maxwell Garnett JC, (1904) Colours in Metal Glasses and in Metallic Films. Philos. Trans. R. Soc. London, 203:385–420.
[9] Looyenga H, (1965) Dielectric constants of heterogeneous mixtures. Physica, 31:401–406.
[10] Bruggemann DAG, (1935), Berechnung Verschiedener Physikalischer Konstanten von Heterogenen Substanzen. Ann. Phys., 24:636.
[11] Setzu S, Lerondel G, Romestain R, (1998) Temperature effect on the roughness of the formation interface of p-type porous silicon. J. Appl. Phys., 84:3129–3134.
[12] Theiß W, (1997) Optical properties of porous silicon. Surf. Sci. Rep., 29:95–192.
[13] Theiß W, Hilbrich S, (1997) Refractive index of porous silicon. In L Canham, editor, Properties of Porous Silicon, volume 18 of Emis Datareviews Series, page 223. INSPEC, IEE, London, United Kingdom.
[14] Aspnes DE, Theeten JB, (1979) Investigation of effective-medium models of microscopic surface-roughness by spectroscopic ellipsometry. Phys. Rev. B, 20:3292.
[15] Squire EK, Snow PA, Russell PS, Canham LT, Simons AJ, Reeves CL, (1998) Light emission from porous silicon single and multiple cavities. J. Lumin., 80:125–128.
[16] Reece PJ, Lerondel G, Zheng WH, Gal M, (2002) Optical microcavities with subnanometer linewidths based on porous silicon. Appl. Phys. Lett., 81:4895–4897.
[17] Kovalev D, Polisski G, Diener J, Heckler H, Künzner N, Timoshenko VYu, Koch F, (2001) Strong in-plane birefringence of spatially nanostructured silicon. Appl. Phys. Lett. 78:916–918.

[18] Diener J, Künzner N, Kovalev D, Gross E, Koch F, (2002) Dichroic behavior of multilayer structures based on anisotropically nanostructured silicon. J. Appl. Phys. 91:6704–6708.
[19] Diener J, Künzner v, Gross E, Kovalev D, Fujii M, (2004) Planar silicon-based light polarizers. Opt. Lett. 29:195–197.
[20] Wu QH, De Silva L, Arnold M, Hodgkinson IJ, Takeuchi E, (2004) All-silicon polarizing filters for near-infrared wavelengths. J. Appl. Phys. 95:402–405.
[21] Bruggemann DAG, (1935), Berechnung Verschiedener Physikalischer Konstanten von Heterogenen Substanzen. Ann. Phys., 24:636.
[22] Zettner J, Thoenissen M, Hierl Th, Brendel R, Schulz M, (1999) Novel porous silicon backside light reflector for thin silicon solar cells. Progress in Photovoltaics: Research and Applications 6:423–432.
[23] Kochergin V, Christophersen M, Föll H, (2004) Effective Medium Approach for Calculations of Optical Anisotropy in Porous Materials. Appl. Phys. B, 79:731–739.
[24] Yaghjian AD, (1980) Electric Dyadic Green's Functions in the Source Region. Proc. IEEE, 68:248–263.
[25] Maldovan M, Bockstaller MR, Thomas EL, Carter WC, (2003) Validation of the effective-medium approximation for the dielectric permittivity of oriented nanoparticle-filled materials: effective permittivity for dielectric nanoparticles in multilayer photonic composites. Appl. Phys. B 76:877–884.
[26] Sihvola A, Lindell IV, (1992) Polarizability Modeling of Heterogeneous Media (In: Progress in Electromagnetics Research, Vol. 6) edited by Priou, A., Elsevier Science Publ Co New York 1992, 101–151.
[27] Cullis AG, Canham LT, Calcott PDJ, (1997) The structural and luminescence properties of porous silicon. J. Appl. Phys., 82:909–912.
[28] Lehmann V, Stengl R, Luigart A, (2000) On the morphology and the electrochemical formation mechanism of mesoporous silicon. Materials science and engineering B, 69:11–22.
[29] Faivre C, Bellet D, (1999) Structural properties of p+-type porous silicon layers versus the substrate orientation: an X-ray diffraction comparative study. J. Appl. Cryst. 32:1134–1144.
[30] Yariv A,Yeh P, Optical Waves in Crystals, Wiley, 1984.
[31] Föll H, Langa S, Carstensen J, Christophersen M, Tiginyanu IM, (2003) Review: Pores in III-V Semiconductors. Adv. Materials, 15:183–198.
[32] Langa S, Tiginyanu IM, Carstensen J, Christophersen M, Föll H, (2000) Formation of porous layers with different morphologies during anodic etching of n-InP. J. Electrochem. Soc. Lett., 3:514–516.
[33] Föll H, Carstensen J, Langa S, Christophersen M,Tiginyanu IM (2003) Porous III-V compound semiconductors: formation, properties and comparison to silicon. Phys. Stat. Sol. A 197:61–70.
[34] Erne BH, Vanmaekelbergh D, Kelly JJ., (1996) Morphology and Strongly Enhanced Photoresponse of GaP Electrodes Made Porous by Anodic Etching. J. Electrochem. Soc., 143:305–314.
[35] Kochergin V, Christophersen M, Föll H, (2005) Adjustable optical anisotropy in porous GaAs. Appl. Phys. Lett., 86:042108.
[36] Langa S, Carstensen J, Christophersen M, Föll H, Tiginyanu IM, (2001) Observation of crossing pores in anodically etched n-GaAs. Appl. Phys. Lett., 78:1074–1076.
[37] Landau LD, Lifshits EM, (1984) Electrodynamics of Continuous Media, 2nd ed. Butterworth-Heinenann, Oxford.
[38] Schmuki P, Erickson LE, (1998) Direct micropatterning of Si and GaAs using electrochemical development of focused ion beam implants. Appl. Phys. Lett., 73:2600–2602.

[39] Schmuki P, Lockwood DJ, Labbe HJ, Fraser JW, (1996) Visible photoluminescence from porous GaAs. Appl. Phys. Lett., 69:1620–1622.
[40] Föll H, Langa S, Carstensen J, Lölkes S, Christophersen M, Tiginyanu IM, (2003) Engineering Porous III-Vs. III-Vs Review, 16:42–43.
[41] Sauer G, Brehm G, Schneider S, Nielsch K, Wehrspohn RB, Choi J, Hofmeister H, Gösele U, (2002) Highly ordered monocrystalline silver nanowire arrays. J. Appl. Phys. 91:3243–3249.
[42] Matthias S, Schilling J, Nielsch K, Müller F, Wehrspohn RB, Gösele U, (2002) Monodisperse Diameter-Modulated Gold Microwires. Adv. Mater. 14:1618–1621.
[43] Kuwata H, Tamaru H, Esumi K, Miyaho K, (2003) Resonant light scattering from metal nanoparticles: Practical analysis beyond Rayleigh approximation. Appl. Phys. Lett. 83:4625–4627.
[44] Moskovits M, (1985) Surface-enhanced spectroscopy. Reviews of Modern Physics, 57:783–826.
[45] Kochergin V, Christophersen M, Föll H, (2005) Surface Plasmon Enhancement of an Optical Anisotropy in Porous Silicon/Metal Composite. Appl. Phys. B., 80:81–87.
[46] (1998) Properties of Porous Silicon Edited by Leigh Canham, IEE Publishing.
[47] Smith DR, Schurig D, (2003) Electromagnetic Wave Propagation in Media with Indefinite Permittivity and Permeability Tensors. Phys. Rev. Lett., 90:77405.

4

Leaky Waveguide Approach of Light Propagation Through Porous Semiconductors

4.1 Introduction

The leaky waveguide mode of light transmission through porous silicon did not gain much attention until recently. This mode of light propagation was first observed by Lehmann et al. [1], who reported the short-wave pass filtering by a macroporous silicon membrane. The experimental findings, however, were only explained by the spectral dependency of light diffraction at a small aperture; which falls somewhat short of the task. Nevertheless, a thin sheet of macroporous material that is opaque at visible wavelengths works principially as a short pass optical filter. A more detailed explanation of this effect in terms of leaky waveguide mode transmission was given in [2] and [3]. It was proven that in pores that are long compared to the wavelength of light waveguiding in the pores becomes the key phenomenon that determines the light transmission in the UV and visible parts of the spectrum. The waveguides, with cores formed by the pores, are leaky since the light can leak out into their higher-index cladding.

This chapter of the book is devoted to the description of the leaky waveguide mode of light propagation through macroporous silicon. The basic description of the leaky waveguides will be given first, and the detailed description of leaky waveguide propagation through macroporous silicon membrane will be provided next. This chapter is nessesary for the full understanding of macroporous silicon-based UV optical filters and polarizers discussed in Chapters 8 and 9.

4.2 Leaky Waveguides

For a normal waveguide, in order to obtain lossless propagation through the dielectric slab, the index of refraction of the waveguiding layer (core) should exceed the refractive indices of the media on both sides of said waveguiding layer. This is a necessary condition for optical waveguiding, although it is not sufficient; the thickness of the waveguiding layer must exceed a cut-off thickness. However, guiding light with very small losses is also possible if the core refractive index is lower than that of at least one medium surrounding the waveguiding layer. In this case, total internal reflection, required in order to suppress the reflection losses completely, cannot be achieved at the interfaces. The light wave coupled into such a structure will thus lose power by "leaking" to the bounding medium with a refractive index exceeding that of the waveguiding layer (or into both media, if their refractive indices exceed those of the waveguiding layer). That is why such a structure is called a leaky waveguide.

Low losses in a leaky waveguide are achieved because the Fresnel reflectivity at the interfaces approaches unity when the angle of incidence reaches 90^0 Leaky waveguide mode characteristics, such as propagation constants (or effective refractive indices) and loss coefficients, can be derived similarly to the case of normal waveguides [4–7]. The TE mode effective refractive index n_{TE}*, for a single-layer leaky waveguide structure with waveguiding layer of index of refraction n_2 having thickness d surrounded by media with refractive indices n_1 and n_3 will be the solution of the analytical equation:

$$\tan(k_{2z}d) = i\frac{k_{2z}(k_{1z}+k_{3z})}{k_{2z}^2 + k_{1z}k_{3z}} \tag{4.1}$$

where $k_{iz} = \frac{\omega}{c}\sqrt{n_i^2 - n_{TE}^{*2}}$, $i = 1,2,3$.

Equation 4.1 is similar to the equation governing normal waveguide mode propagation [5]. The difference is due to the complex nature of the leaky mode propagation constant: $\beta = (\omega/c)\mathrm{Re}(n_{TE}^*) - i\alpha/2$, where $\alpha > 0$ is the *mode power attenuation coefficient*, also known as the *mode loss coefficient*. The complex propagation constant β corresponds to an exponential decay of the leaky waveguide mode power along the mode propagation direction.

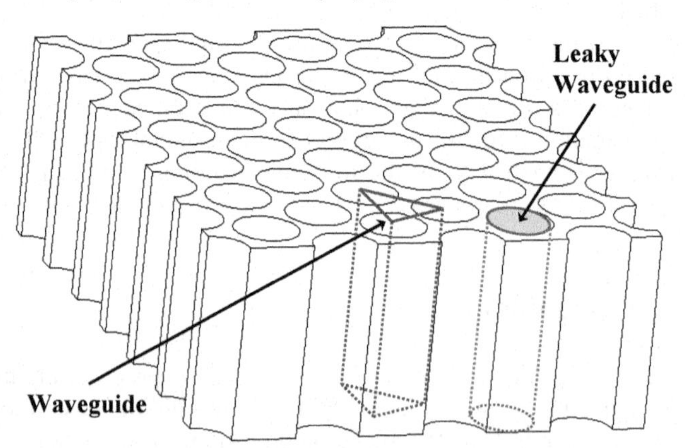

Figure 4.1. The structure of short-pass MPSi filter structure (Lehmann's filter)

In general, there is no analytical solution of (4.1). It can be solved approximately (see [5]) or numerically. Similarly to normal waveguides, leaky waveguide modes have cut-off conditions (defined by wavelenth, geometrical and optical parameters of leaky waveguide structure). In general, the leakage losses of leaky waveguide modes increase exponentially near the leaky waveguide mode cut-off conditions (although for certain multilayer structures this may not be the case). Despite the fact that leaky waveguides are lossy by definition, they found a number of applications in areas where very long loss-lesss propagation is not nessesary (such as integrated optics on a silicon chip, where propagation through centitimeters of waveguides is sufficient).

4.3 Leaky Waveguide Transmission Through MPSi Array

As mentioned previously, it was Lehmann et al. [1] who first observed the UV channel of transmission through macroporous silicon array (that will be denoted in the future consideration as *Lehmann's filter*). The structure of Lehmann's filter is schematically shown in Figure 4.1, while the measured transmission spectra through Lehmann filters with different macroporous silicon parameters are given in Figure 4.2. Such a filter consists of air- or vacuum-filled macropores electro-etched into the silicon wafer host. The macropores are forming an ordered uniform array (in [1] the array was of cubic symmetry). The pore's ends are open on both the first and second interfaces of silicon wafer, the confguration is that of a membrane

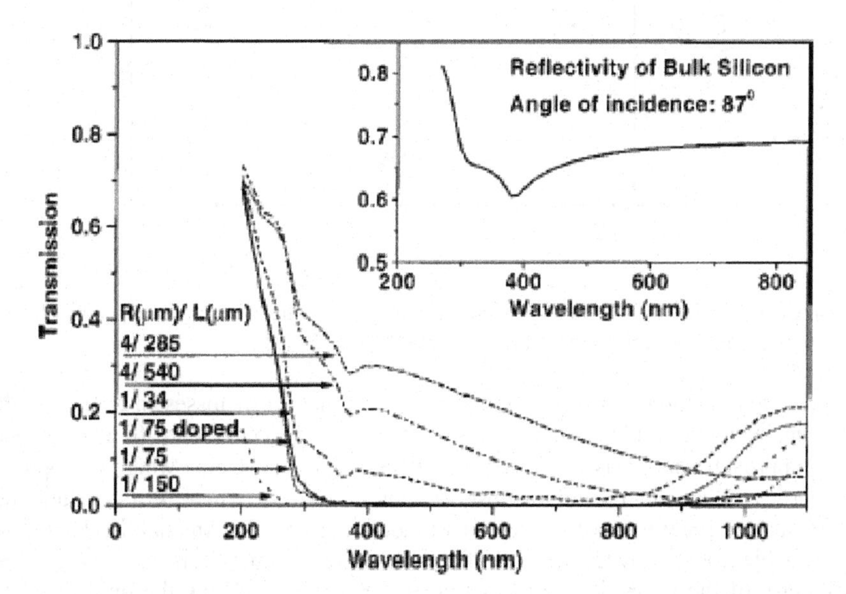

Figure 4.2. Measured transmittance (corrected for porosity) of MPSi array with pore radius R, pore length L, and the pore axis parallel to the light beam. Inset shows the spectral reflectivity of bulk silicon for a large angle of incidence (After [1])

In the UV and visible spectral range the transmission through such an MPSi array obviously takes place through the pores (since silicon is opaque at these wavelengths, as follows from Figure 4.3). Each pore in such a material can be considered as a waveguide. The waveguides, with cores formed by the pores, are essentially leaky since the Si cladding has a relatively large index of refraction. Optical filtering in these long pores then is mainly provided by wavelength-dependent propagation losses. Using a simplified geometrical picture, a short-wavelength fundamental mode in a pore waveguide is presented by a ray trajectory with a shallow angle almost parallel to the pore walls, while a trajectory corresponding to long-wavelength mode would make a larger angle with respect to the pore walls. This causes more reflection events per unit of length as well as a smaller reflection, which results in higher optical losses for longer wavelengths.

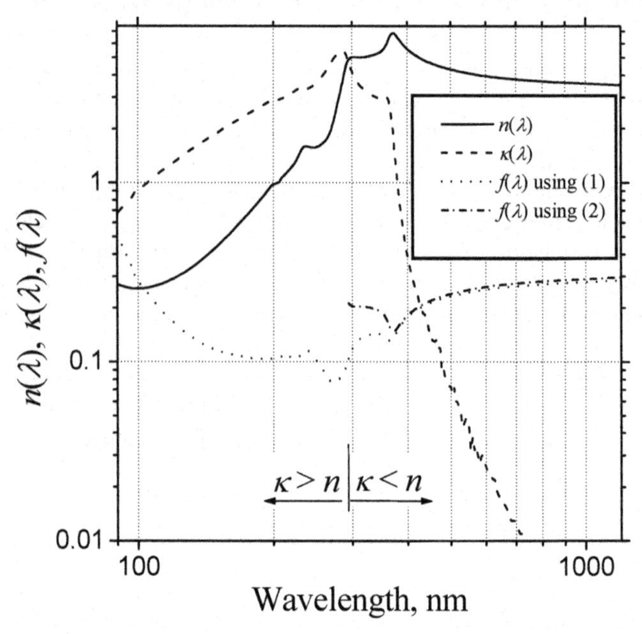

Figure 4.3. The optical constants of silicon (from [8]) and the wavelength-dependent factor f(λ) for optical losses calculated using (4.2) and (4.3) (After [3])

This explanation can be supported by an estimation of losses in a waveguide formed by metallic-like boundaries surrounding a dielectric core (e.g. vacuum). This approach can be used at short wavelengths ($\lambda < 294$ nm) where the optical properties of silicon resemble those of metals ($\kappa > n$, where κ and n are the imaginary and real parts of the refractive index, Figure 4.3). An analytical solution is possible for a waveguide with flat boundaries. While it is not exactly the geometry of the pores, it gives a reasonable estimation of how the optical losses depend on the wavelength. For the N-th order ($N = 0,1,2,\ldots$) transverse electric (TE) mode in a vacuum slab between the metal mirrors ($\kappa \gg n$, $\kappa \gg 1$) separated by a distance $d \gg \lambda$ this estimation is (after [9]):

$$\alpha_{metal}^{TE} \approx (N+1)^2 \frac{\lambda^2}{d^3} \frac{n}{n^2+\kappa^2} \tag{4.2}$$

In the wavelength range $\lambda > 294$ nm the real part of the index is greater than the imaginary part, $n > \kappa$. Waveguiding in the pores is still provided by the same mechanism, high reflection for a shallow incident angle. The difference to the metallic mirrors is in the electromagnetic field structure in the waveguide claddings. In the case of metallic claddings, the field is strongly decaying while in the high index dielectric cladding, the waveguide mode looses energy to waves propagating in the cladding, i.e. the waveguide becomes leaky. Due to the high absorption of silicon in this spectral range, the leaking waves are eventually absorbed in any case. Optical losses due to leakage only ($\kappa \ll n$, $d \gg \lambda$) are estimated as follows (after [5]):

$$\alpha_{leakage}^{TE} \approx (N+1)^2 \frac{\lambda^2}{d^3} \frac{1}{\sqrt{n^2-1}} \tag{4.3}$$

Both equations (4.2) and (4.3) predict similar dependence of optical losses on wavelength, waveguide thickness, and the mode order. Moreover, Equations 4.2 and 4.3 predict almost identical values of losses for $\lambda > 380$ nm. Hence, Equation 4.2 can be used over the entire spectrum for practical estimations. The transmission through the waveguide of length L is proportional to an exponential factor such as $exp(-\alpha L)$, which transfers a quadratic dependence of losses versus wavelength into a steep function with strong filtering of light with longer wavelengths. The macroporous material becomes a short-pass filter. However, at wavelengths $\lambda >$ 1100 nm silicon is transparent, resulting in another channel of transmission through macroporous silicon, which will be considered in detail in Chapter 5. In fact, as it is shown below, absorption in silicon becomes lower than the leakage losses at approximately $\lambda \approx 700$–800 nm, which further restricts the applicability of the simple analysis presented in this paragraph.

For simplified estimations we note that for a particular material, silicon, the factors in (4.2) and (4.3) depend on the optical indices and do not change considerably across the UV and visible spectrum. The index factor

$$f(\lambda) = \begin{cases} n(\lambda)/\left(n(\lambda)^2 + \kappa(\lambda)^2\right) & \text{if} \quad n(\lambda) < \kappa(\lambda) \\ 1/\sqrt{n(\lambda)^2 - 1} & \text{if} \quad n(\lambda) > \kappa(\lambda) \end{cases} \tag{4.4}$$

calculated using published data for the optical constants of silicon is shown in Figure 4.3. Note that (4.2) and (4.3) are valid if $n \ll \kappa$ and $n \gg \kappa$ respectively, which may result in a discontinuity in (4.4) when $n \approx \kappa$. The function $f(\lambda)$ does not change much across the entire spectral range of interest. Moreover, only the fundamental mode ($N = 0$) is essential because of the rapid increase of losses with N increasing. As long as the excitation of odd modes is unlikely due to a vanishing overlapping integral of the incident flat wave and the modal field, the next mode after the fundamental is the mode with $N = 2$. Losses for this mode are almost one order of magnitude ($(N + 1)^2 = 9$) higher. This reduces (4.2) and (4.3) to the very simple and practical formula

$$\alpha = f_0 \lambda^2 / d^3 . \tag{4.5}$$

By comparing losses calculated using (4.5) (at the assumption of $f_0 \approx 0.09$) with exact numerical calculations for a slab waveguide, one can find that (4.5) gives a reasonably good (± 10% accuracy) estimation in the 90–300 nm wavelength range, despite of the fact that at the short wavelengths the condition $\kappa >> 1$ is violated (Figure 4.4). In the visible range, exact calculations give slightly higher losses. For wavelengths longer than 500 nm, Equation 4.5 must be used with $f_0 \approx 0.28$.

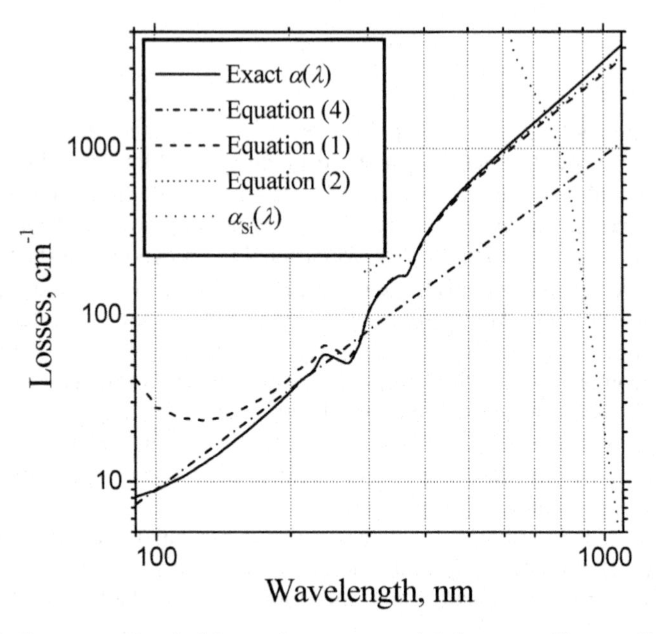

Figure 4.4. Spectrum of optical losses in a vacuum slab between silicon walls calculated using exact numerical algorithm and estimated using (4.2) and (4.3), and (4.5). After [3]

Besides the leaky wavegide mode losses, the leaky waveguide transmission through the macroporous silicon layer is also defined by the light coupling to the pore waveguides. A simple estimation can be done for pores with rectangular ($a \times b$) cross section. In the deep UV range, where optical constants of silicon are almost metallic, we can assume that the modal field has a sinusoidal profile:

$$E_{N,M}(x, y) = \sin((N+1)\pi x / a)\sin((M+1)\pi y / b) \tag{4.6}$$

where N, M = 0, 1, 2, 3... are the mode order indices. The beginning of the Cartesian coordinate system is placed at the corner of the rectangular pore. The squared overlapping integral of this field with the plane incident wave becomes equal to:

$$I_{N,M} = \frac{64}{\pi^4} \frac{1}{(N+1)^2} \frac{1}{(M+1)^2} \tag{4.7}$$

if N and M are even; 0 otherwise.

Thus, in addition to the porosity factor determined as the relative area covered by pores, one has to apply an additional coupling factor for the fundamental mode equal to $64/\pi^4 \approx 0.66$. High order modes have much smaller coupling: e.g. $I_{2,0} = I_{0,2} = (1/9)I_{0,0}$; $I_{4,0} = I_{0,4} = (1/25)I_{0,0}$; $I_{2,2} = (1/81)I_{0,0}$ etc. They also experience stronger attenuation (see (4.2) and (4.3)). Total coupling to all the modes in a very wide pore supporting a large number of modes equals 100% (if not counting the porosity factor) due to a mathematical identity

$$\sum_{p=0}^{\infty} \sum_{q=0}^{\infty} \frac{1}{(2p+1)^2} \frac{1}{(2q+1)^2} = \frac{\pi^4}{64} \tag{4.8}$$

As mentioned previously, and as will be considered in detail in Chapter 5, in the near IR and IR wavelength range, the nature of the transmission through the MPSi layer changes and waveguide mode of transmission through silicon islands between the pores becomes significant.

Figure 4.5 shows the calculated spectral dependences of the optical fundamental leaky waveguide and of the waveguide modes loss coefficients for $1 \times 1\ \mu m^2$ near-square vacuum-filled pores in silicon. According to Figure 4.5, transmission through the pores by the leaky waveguide mechanism is dominant up to ~700 nm, the transmission through the silicon host waveguides is dominant starting from ~800 nm, while at 700 to 800 nm both transmission mechanisms compete with each other.

In Figure 4.6 the plot of the spectral dependence of the transmission through an MPSi array of cubic symmetry, 50 μm thickness, and 1 μm pore diameters, is given. The curves were calculated according to formalism presented above. The close matching with the experimental data of [1], reproduced in Figure 4.2, validates the model discussed above.

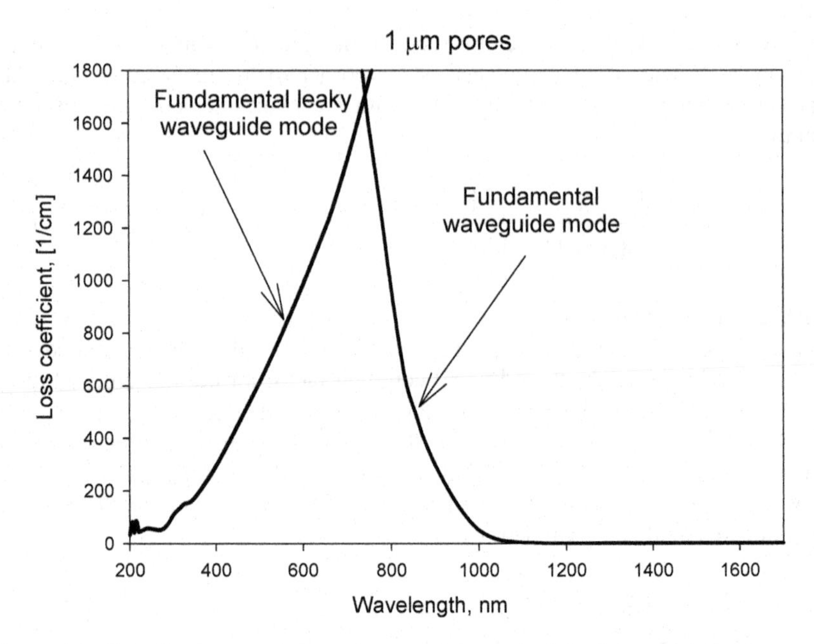

Figure 4.5. Calculated spectral dependences of the optical fundamental leaky waveguide and waveguide modes loss coefficients for 1×1 μm^2 near-square pore in silicon. After [2]

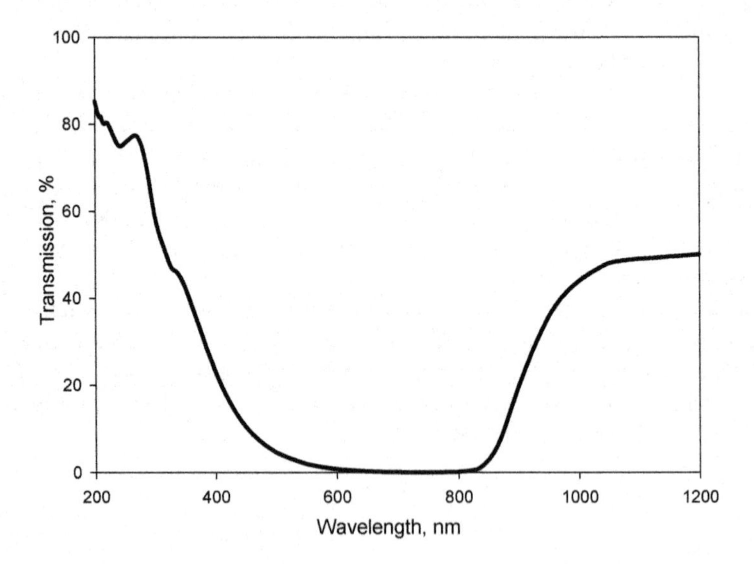

Figure 4.6. The numerically calculated spectral dependence of transmission through MPSi array of cubic symmetry, 50 μm thickness, and 1 μm pore diameters. After [2]

4.4 Conclusions

Leaky waveguide mechanism of transmission through MPSi arrays, while initially somewhat unexpected and discovered relatively late, opens up a number of potential optical applications of macroporous silicon, which will be reviewed in more detail in Chapters 8 and 9. This mechanism can also be expected in other porous semiconductors with straight pores such as, e.g. porous InP [10]. While porous alumina is not expected to provide efficient leaky waveguide transmission channels due to the low reflectivity of alumina, it might provide efficient UV leaky waveguide transmission channel if the pores can be conformally coated with metal a layer,. We believe that more unexpected applications of this unique mode of transmission through porous semiconductors will be found in the future, since this mode of transmission provides engineers and researchers with additional degree of freedom in designing the optical properties of materials.

4.5 References

[1] Lehmann V, Stengl R, Reisinger H, Detemple R, Theiss W, (2001) Optical shortpassfilters based on macroporous silicon. Appl. Phys. Lett. 78:589–591.
[2] Kochergin V, (2003) Omnidirectional Optical Filters. Kluwer Academic Publishers, Boston, ISBN 1-4020-7386-0.
[3] Avrutsky I, Kochergin V, (2003) Filtering by leaky guided modes in macroporous silicon. Appl. Phys. Lett. 82: 3590–3592.
[4] Marcuse D, (1974) Theory of Dielectric Optical Waveguides, Academic Press.
[5] Yariv A, Yeh P, (1984) Optical Waves in Crystals: Propagation and Control of Laser Radiation. John Wiley & Sons.
[6] Yeh P, (1988) Optical Waves in Layered Media. John Wiley & Sons.
[7] Okamoto K, (2000) Fundamentals of Optical Waveguides, Academic Press.
[8] Palik ED, (1988) Handbook of Optical Constants of Solids, Academic Press.
[9] Adams MJ, (1981) An Introduction to Optical Waveguides. Wiley.
[10] Föll H, Langa S, Carstensen J, Christophersen M, Tiginyanu IM, (2003) Review: Pores in III-V Semiconductors. Adv. Materials, 15:183–198.

5

Normal Waveguide Approach of Light Propagation Through Porous Semiconductors

5.1 Introduction

The waveguide mode of light transmission through porous silicon was discovered later than other modes of transmission, probably because it appears only in certain geometries of macroporous silicon layer. Lehmann et al. [1], who reported the presence of an IR transmission band through the ordered macroporous silicon array can be credited with the first observation of this mode of light propagation. No explanation of the experimental findings (IR channel of transmission) was suggested though, since the contribution was mainly devoted to the evaluation of the UV short pass behavior of the macroporous silicon array. The existence of a waveguide mode of transmission through MPSi arrays was first predicted in [2]. It was shown that in the IR range at certain geometries of the MPSi layer the waveguiding in the silicon cores between pores becomes the key phenomenon that determines the light transmission. It was also suggested in [2] that the waveguide mode of transmission through MPSi layers can enable a new class of optical components and devices, namely, omnidirectional band-pass and narrowband pass filters, potentially important for a number of applications.

This chapter is addressing the waveguide mode of light propagation through macroporous silicon. The basic description of the waveguides is provided first, and a detailed description of the waveguide propagation through a macroporous silicon layer is provided next. This chapter is nessesary for the understanding of macroporous silicon-based omnidirectional filters for the IR range, discussed in detail in Chapter 11.

5.2 Dielectric Waveguides

In general, waveguide is a structure that supports confined propagation of electromagnetic waves in a waveguide mode [3–6]. A large number of different waveguiding structures are used today in telecommunication, sensing and other applications. The simplest waveguide structure is the single-layer plane dielectric waveguide. It consists of a thin dielectric layer or core sandwiched between two semi-infinite bounding media (substrate and cladding). The index of refraction of the waveguiding dielectric layer has to be greater than those of the surrounding media to ensure total internal reflection of the light propagating in the dielectric at both interfaces..

A waveguide mode can be written in the form $\boldsymbol{E}^w(\boldsymbol{r}, t) = \boldsymbol{E}_0^w(\boldsymbol{r})\cdot\exp[i(\omega\cdot t-(\omega/c)\cdot n^*\cdot z)]$, where z is the coordinate along the waveguide propagation axis, n^* is the *effective refractive index* of the waveguide mode, and $\boldsymbol{E}_0^w(\boldsymbol{r})$ is the electric field distribution of the mode across the waveguide cross-section, which should be constant along the direction of mode propagation in the uniform waveguide. In general, a waveguide structure supports multiple propagation modes with different effective refractive indices and different electric field distributions. For the case of one.dimensional waveguide structures with a refractive indices n_1, n_2 and n_0 of the core, the cladding and the substrate, respectively, the effective refractive indices of waveguide modes for TE polarization are solutions of the following equation:

$$\tan(hd) = \frac{p+q}{h(1-pq/h^2)} \tag{5.1}$$

where $h = (\omega/c)[n_1^2-(n^*)^2]^{1/2}$; $q = (\omega/c)[(n^*)^2-n_0^2]^{1/2}$; $p = (\omega/c)[(n^*)^2-n_2^2]^{1/2}$ and d = thickness of the core. For TM modes one finds

$$\tan(hd) = \frac{h(p'+q')}{h^2 - p'q'} \tag{5.2}$$

where $p' = (n_1/n_2)^2 p$ and $q' = (n_1/n_0)^2 q$.

A waveguide mode appears (i.e. the light wave becomes confined) above a certain value of d/λ, which is called *mode cutoff*. At or below the cutoff value, the mode extends to $x = \pm\infty$ i.e., the mode becomes unconfined and cannot be called a waveguide mode anymore. This means that for any given waveguide structure there is a certain maxumum wavelength for each mode, above which this mode vanishes. In reverse, for any given wavelength there is a smallest waveguiding layer dimensions below which this mode does not exist. For a one dimensional waveguide the cutoff values of d/λ are given by

$$\left(\frac{d}{\lambda}\right)_{TE} = \frac{1}{2\pi\sqrt{n_1^2 - n_2^2}}\left[m\pi + \tan^{-1}\left(\frac{n_2^2 - n_0^2}{n_1^2 - n_2^2}\right)^{1/2}\right] \tag{5.3a}$$

$$\left(\frac{d}{\lambda}\right)_{TM} = \frac{1}{2\pi\sqrt{n_1^2 - n_2^2}}\left[m\pi + \tan^{-1}\frac{n_1^2}{n_0^2}\left(\frac{n_2^2 - n_0^2}{n_1^2 - n_2^2}\right)^{1/2}\right] \tag{5.3b}$$

where $m = 0, 1, 2, 3,\ldots$ corresponds to the waveguide mode numbers (i.e., TE_m or TM_m). As follows from (5.3), the number of modes supported by the waveguide structure with a given ratio d/λ depends on the refractive index contrasts between the waveguiding layer and the surrounding layers. In general, the higher the contrasts, the more waveguide modes exist for the given value of d/λ or, reciprocally, a smaller thickness of the waveguiding layer is required for the given wavelength to support a given number of waveguide modes.

For any given waveguide structure and given wavelength, the waveguide modes that correspond to this structure have a distinct electric field distribution. Confinement of the waveguide modes is decreased with the mode number and the confinement of the TM mode is generally weaker than the confinement of the TE mode for the same mode number. Another important property of waveguide modes is a higher mode field localization in high refractive index layers compared to lower refractive index layers. Consequently, the waveguide mode electric (and magnetic) field distribution is asymmetric in an asymmetric waveguide structure and symmetric in a symmetric waveguide structure.

An important property of waveguide modes is their orthogonality: $C_{lm}^{TE} \int \boldsymbol{E}_l(\boldsymbol{r}) \cdot \boldsymbol{E}_m^*(\boldsymbol{r})\, \mathrm{d}\, x = \delta_{lm}$; $C_{lm}^{TM} \int \boldsymbol{H}_l(\boldsymbol{r}) \cdot \boldsymbol{H}_m^*(\boldsymbol{r})\, \mathrm{d}\, \boldsymbol{r} = \delta_{lm}$, where C_{lm}^{TE} and C_{lm}^{TM} are some constants defined by the normalization of modes and δ_{lm} is the Kronecker delta. In addition, for the waveguide structures consisting of isotropic materials, all the TE and TM modes are mutually orthogonal: $C_{lm} \int (\boldsymbol{E}_l(\boldsymbol{r})\ \mathbf{x}\ \boldsymbol{H}_m^*(\boldsymbol{r})) \cdot \mathbf{z}\, \mathrm{d}\, \boldsymbol{r} = \delta_{lm}$, where C_{lm} is some constant, defined by normalization, z is the unit vector in the mode propagation direction and $\boldsymbol{r}$ is the vector in the plane perpendicular to z. It means that for any given waveguide structure and any given wavelength, there is one and only one set of waveguide modes that is supported by this structure, and that any light wave, propagating through the waveguide structure, can be expanded into the set of independent modes with unique relative power weights.

An important parameter of optical waveguides is the *numerical aperture* (*NA*). NA is defined as the largest angle θ_{max} of incident rays with the normal that experiences total internal reflection (TIR) in the waveguiding layer (or the core). For a symmetrical one-dimensional waveguide i.e., when $n_0 = n_2$

$$NA = \sin\theta_{max} \approx [(n_1)^2 - (n_0)^2]^{1/2}\,. \tag{5.4}$$

Rays launched outside θ_{max}, will not excite the waveguide mode. If we define the relative refractive index difference between core and cladding in a symmetrical waveguide as $\Delta = (n_1 - n_0)/n_1$, then the NA is related to Δ as $NA = \sin\theta_{max} \approx n_1(2\Delta)^{1/2}$. The maximum angle for the propagating light within the core (or waveguiding layer) is $\sin\varphi_{max} \approx \theta_{max}/n_1$.

Another important phenomenon that we will need for the future consideration is coupling between adjacent waveguides. The effect can be easily understood on a one-dimensional example of two planar waveguides, spaced apart by some distance. For the sake of simplicity, both waveguides can be assumed to be single-mode. In the case of waveguides being placed sufficiently close together (so the overlap of the waveguide modes exists), the mode splitting takes place into even (first-order mode in the multilayer waveguide, obtained by placing two initial waveguides close together) and odd (second-order mode) hybrid modes of the structure. If the incident electromagnetic wave is coupled to one of the waveguides, it experieces a complete shift to the other waveguide at the distance $L_c = \lambda /2(n_e^* - n_o^*)$, where n_e^* and n_o^* are effective refractive indices of even and odd modes, respectively. Due to the time invariance of the Maxwell equations, the result will be the same if the incident field is coupled to the other waveguide. In general, the coupling length increases exponentially with the waveguide core separation distance.

5.3 Waveguide Transmission Through MPSi Array

Waveguide mode propagation through porous silicon is possible at certain geometries of macroporous silicon arrays along the direction of pore growth. Let's consider a system that is made of an array of waveguides made in the form of a macroporous silicon (MPSi) layer where silicon (refractive index n_{Si} = 3.5) cores between the pores serve as waveguide cores above the silicon absorption edge, i.e., at λ > 1100 nm, while the air-filled pores (n_{Air} = 1) serve to optically decouple these waveguides. Such arrays are schematically shown in Figure 5.1. Three physical processes affect the transmission through such a system: coupling of light into the structure at the first MPSi layer interface, propagation of light through the structure, and outcoupling at the second MPSi structure interface.

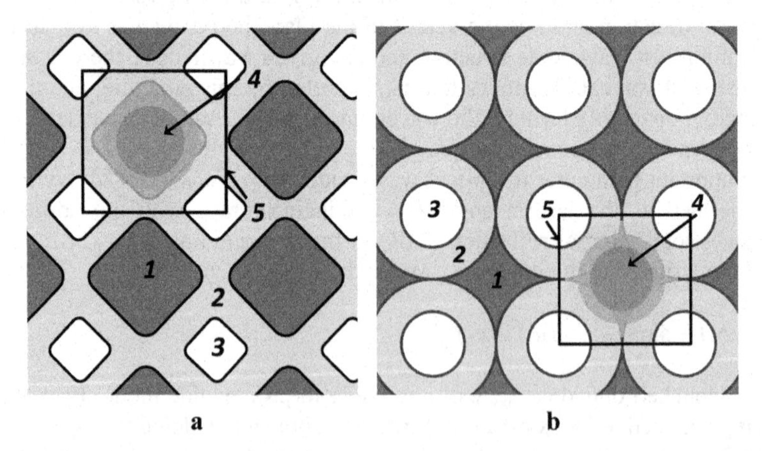

Figure 5.1. Drawings of the waveguide arrays inherent to ordered macroporous silicon. The silicon cores (1), oxide layers (2), pores (3), waveguide mode volume (4) and the array's elementary cells (5) are shown. Possible realizations of cubic-symmetry MPSi arays are illustrated: **a** cubic symmetry, near-square pores; **b** cubic symmetry, circular pores. After [2]

The efficiency $P_{i,j}^{W}(\lambda)$of the coupling of normally incident light at the first MPSi layer interface into the MPSi structure for the $(i,j)^{th}$ mode can be calculated as

$$P_{i,j}^{W}(\lambda)=\frac{\int E_{i,j}^{W}(s,\lambda)\cdot E_{I,j}^{*}(s,\lambda)\cdot ds}{\sqrt{\int E_{i,j}^{W}(s,\lambda)\cdot(E_{i,j}^{W})^{*}(s,\lambda)\cdot ds\cdot\int E_{I,j}(s,\lambda)\cdot E_{I,j}^{*}(s,\lambda)\cdot ds}} \tag{5.6}$$

where $E_{i,j}^{W}$ (s,λ) is the electric field of an $(i,j)^{th}$ waveguide mode at the plane of the first MPSi layer interface; E_I(s, λ) is the electric field of the incident wave on the MPSi layer perpendicular to the mode propagation direction; and λ is the wavelength of the light.

If a plane-parallel beam of light is normally incident on the first MPSi layer interface, for the fundamental waveguide mode equation (5.6) can be roughly estimated as:

$$P_{0,0}^{W}(\lambda)\approx\frac{4n_{Si}(\lambda)\cdot n_I}{(n_{Si}(\lambda)+n_I)^2}\cdot\frac{S_{uc}-S_p}{S_{uc}} \tag{5.7}$$

where S_p is the area of each pore (see Figure 5.1), S_{uc} is the area of an MPSi array's unit cell (which can be introduced for ordered MPSi arrays only), $n_{Si}(\lambda)$ is the refractive index of the silicon at the wavelength λ, and n_I is the refractive index of the incident medium. S_p for a circular pore cross-section is equal to $\pi\, ø^2/4$, where $ø$ is the diameter of the pore, and for near-square pores the cross-section is ~ d^2, where d is the characteristic cross-section size. For the most common case of air, expression (5.7) can be rewritten as

$$P_{0,0}^{W}(\lambda)\approx\frac{4n_{Si}(\lambda)}{(n_{Si}(\lambda)+1)^2}\cdot\frac{S_{uc}-S_p}{S_{uc}} \tag{5.8}$$

In other words, to some approximation, $P_{0,0}^{W}(\lambda)\approx 0.69(1-p)$, where p is the porosity of the MPSi layer. It should be noted that the cross-coupling between neighbouring waveguides is so far ignored. If cross-coupling is significant Equation 5.8 has a different form.

At the second surface of the MPSi layer, the light from each end of each waveguide is emitted with a divergence governed by the numerical aperture of the waveguide, *NA*, and the wavelength. In the far field, destructive and constructive interference from all light sources, i.e the end of the wave guides, takes place. In the case of an ordered MPSi array, this leads to a number of diffraction orders, and this number is in general defined by the pore array geometry (by the relationship between pore size and pore-to-pore distance) and the wavelength of light. For the majority of optical applications the macroporous silicon component will be positioned in the far field of the receiver (or light coupling device). Therefore, only the 0th diffraction order, the efficiency of which is denoted in the following discussion as $DE_{0,0}^{W}(\lambda)$, is of interest. The only application that is not sensitive to the outcoupling of light to higher diffraction orders demands that the macroporous silicon component is mounted directly on top of photodetector (i.e., the receiver is in the near field of a component). In all other cases, the outcoupling losses at the second MPSi layer interface can be estimated as $1-DE_{0,0}^{W}(\lambda)$ for the waveguide array, since the main source of such losses is the redistribution of light into higher diffraction orders. Such losses are sensitive to both wavelength and pore array geometry. They are more pronounced at short wavelengths due to the higher number of diffraction orders.

The optical losses $\exp(-\alpha_{i,j}^{W}(\lambda)\, l)$ of the normal waveguide mode (i.e. during the propagation of light through the structure) are small in the mid IR range if no absorptive layer or absorptive filling is used in the pores. This is due to the high transparency of the silicon typically used for macroporous silicon fabrication. However, waveguides are known to exhibit the Bragg reflection phenomenon when their effective refractive indices are periodically modulated in the direction of the waveguide mode propagation. Modulation of effective refractive indices in silicon core waveguides in the MPSi layer is possible through modulation of the porosity (or, in other words, modulation of the macropore diameter) of the MPSi layer with its depth. Different ways to fabricate such MPSi layers are discussed in Chapter 11. The modulation of the MPSi layer porosity will modify the spectral dependence of the transmission through each silicon core waveguide thus allowing to design new optical components. This modification can be taken into account by introducing an additional term $T_{i,j}^{BG}$ in expression (2.1):

$$T(\lambda,\theta,\theta') = \Sigma\, DE_{i,j}^{W}(\lambda,\theta')\, P_{i,j}^{W}(\lambda,\theta)\, T_{i,j}^{BG} \cdot \exp(-\alpha_{i,j}^{W}(\lambda)\, l) \tag{5.9}$$

Since Bragg reflections take place during waveguide mode propagation through the waveguide, under the assumption of the independence (i.e., low cross coupling) of waveguides in the waveguide array, this process is independent of the angle of incidence. Hence, if the Bragg grating part $T_{i,j}^{BG}$ of (5.9) contributes most to the overall transmission spectral shape of the MPSi layer, this shape will be independent of the angle of incidence.

The particular shape of the spectral dependence of $T_{i,j}^{BG}$ is a function of the silicon core shape size, wavelength, waveguide mode losses, and so on. It means that the uniform silicon cross-sections, core sizes, and shapes are required. The spectral dependence of $T_{i,j}^{BG}$ is different for different orders and polarizations of waveguide modes. Hence, silicon cores should act as single-mode waveguides at wavelengths for which the optical component is designed. Single mode requirement is also important to ensure independence of the transmission spectral shape on the angle of incidence, due to different dependences of the coupling and outcoupling efficiencies for different-order waveguide modes. Unlike leaky waveguides discussed previously in this book, where losses dramatically increase with the number of modes, the losses of normal (nonleaky) waveguide modes are generally small, so in the case of the multimode waveguiding, higher-order mode transmission cannot be neglected.

To summarize, for the MPSi array to act as an omnidirectional optical component, it should be well-ordered, and the silicon core waveguides should be independent (low cross-coupling) and single-mode. Other than waveguide modes of light transmission through such MPSi arrays should also be suppressed. In this case the expression (5.9) for the 0th diffraction order will take the form

$$T(\lambda,\theta) = DE_{0,0}{}^{W}(\lambda,\theta')\ P_{0,0}{}^{W}(\lambda,\theta)\ T_{0,0}{}^{BG} \exp(-\alpha_{0,0}{}^{W}(\lambda)\ l) \tag{5.10}$$

since for the ordered array of waveguides and zero diffraction order $\theta = \theta'$.

The coupling efficiency at the first MPSi layer surface can be estimated as

$$P_{0,0}{}^{W}(\lambda,\theta) = P_{0,0}{}^{W}(\lambda,\theta = 0) \exp(-(\theta/\theta_{ac})^2) \tag{5.11}$$

where $P_{0,0}{}^{W}(\lambda,\theta = 0)$ is the coupling efficiency into the waveguide array for a plane-parallel beam incident normally to the MPSi layer (which can be estimated according to Equation 5.8), and θ_{ac} is the acceptance angle of the silicon core waveguide. It should be noted that the porosity (i.e., macropore diameters) may not be constant across the MPSi layer. Hence, (5.11) can be rewritten as

$$P_{0,0}{}^{W}(\lambda,\theta) = 0.69(1-p(0)) \exp(-(\theta/\theta_{ac})^2) \tag{5.12}$$

As follows from Equation 5.12 the coupling losses can be minimized by minimizing the porosity at the surfaces of the MPSi layer.

The particular value of the acceptance angle strongly depends on the waveguide structure. For silicon core waveguides and air-filled pores, the $\sin\theta_{ac}$ estimation will be $[(n_{Si})^2-1]^{1/2} = 3.35$; that is the acceptance angle of the silicon core waveguide (and through that of the MPSi layer as a waveguide array) will be $\pi/2$. Such a wide acceptance range of the MPSi layer at IR wavelength rangea means that optical components based on the MPSi layer in the waveguide transmission mode will be truly omnidirectional.

As mentioned above, the absence of the cross-coupling between neighboring silicon core waveguides of the MPSi layer is essential. However, the calculations indicate that for "pure-silicon" MPSi layers the cross coupling is too strong. The simplest way to suppress the cross-coupling is to partially oxidize the MPSi layer so that the silicon core waveguides will be physically separated by either air pores or silicon dioxide bridges. This is illustrated in Figure 5.1, where a thermally grown SiO_2 layer (5) has been included shown.

Before estimating cross-coupling between neighboring waveguides, let us first find the silicon core size and the silicon dioxide layer thickness ranges needed to guarantee the single-mode character of the silicon core waveguides.

Figure 5.2 shows the calculated dependences of the effective refractive indices of the TE polarization waveguide modes on the silicon core cross-section for the cubic-symmetry array of nearly square pores. The wavelength of light was assumed to be 1550 nm. One can see that the silicon core waveguides become multimode starting at about 300 nm of silicon core cross-section. For other wavelengths the maximum silicon core cross-section needed can be found by scaling.

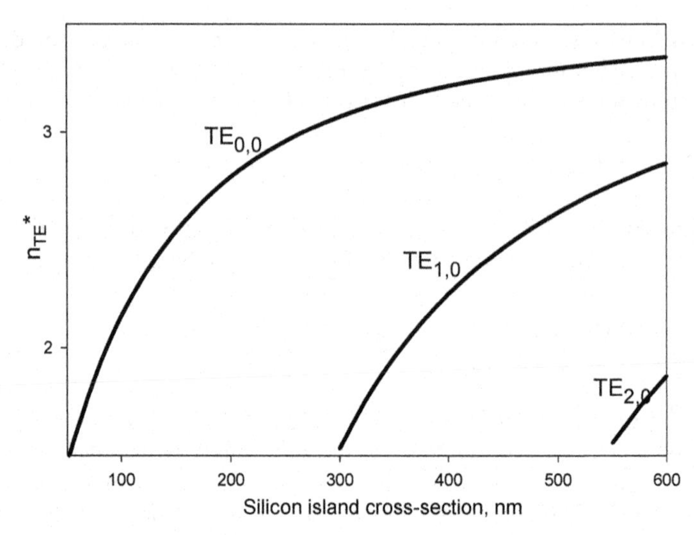

Figure 5.2. Calculated dependences of effective refractive indices of TE polarization waveguide modes on the square silicon core cross-section (side of the square). After [2]

The SiO_2 layer thickness affects both the cross-coupling suppression and the number of modes in the silicon core waveguide for constant silicon core cross-section. In Figure 5.3 the calculated dependences of effective refractive indices of TE polarization waveguide modes on the silicon dioxide layer thickness are given for the MPSi layer of Figure 5.1a. The wavelength of light was assumed to be 1550 nm, while the silicon core cross-section was assumed to be 275 nm. One can see that even for such a silicon core cross-section the thickness of the silicon dioxide layer covering pore walls has to be less than approximately 300 nm.

Now having the silicon core size, the silicon dioxide layer thickness, and the geometry of the MPSi array, we can estimate the cross-coupling coefficient between neighbor waveguides. Figure 5.4 gives the calculated dependence of the cross-coupling coefficient on the silicon core separation is given for fundamental waveguide mode. The wavelength was assumed to be 1550 nm, the silicon core cross-section was assumed to be 275 nm, the silicon dioxide layer thickness was assumed to be 300 nm, and the MPSi layer structure was assumed to be the same as in Figure 5.1a. As expected, the coupling coefficient decreases exponentially as the silicon core spacing decreases. To provide omnidirectionality of the transmitted light, the coupling length, which is equal to the inverse coupling coefficient, should be less than the thickness of the MPSi layer. Taking into account that the typical MPSi layer thickness is in the range of 50 to 300 μm, while the thickness of the MPSi layer with modulated pores diameters will not exceed 200 μm, it will be safe to set the upper limit of the coupling coefficient at about 20 cm^{-1}.

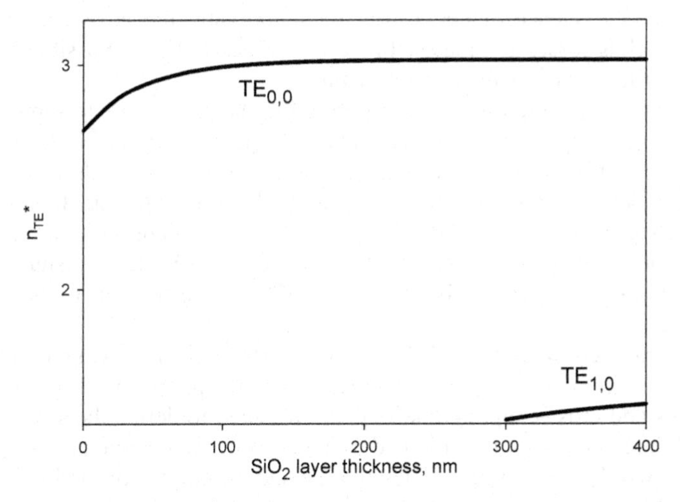

Figure 5.3. Calculated dependences of effective refractive indices of TE polarization waveguide modes on the SiO_2 layer thickness for the structure of Figure 5.1a. After [2]

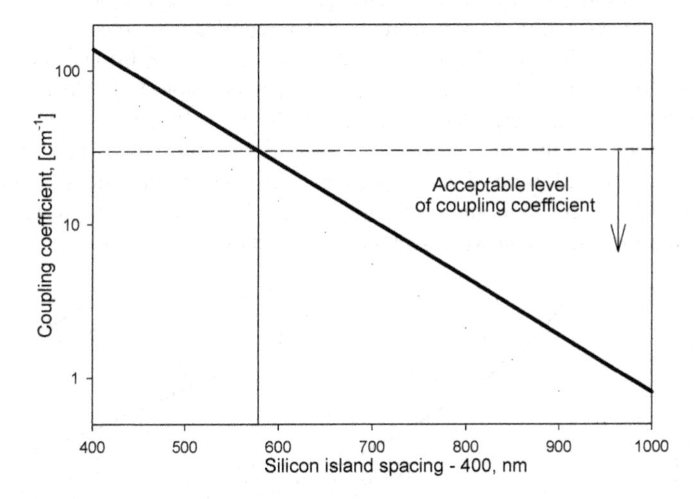

Figure 5.4. Calculated dependences of cross-coupling coefficients of fundamental waveguide modes on the silicon core separation. After [2]

According to Figure 5.4, the cross-coupling coefficient will be suitable for providing omnidirectionality of the transmittance spectrum for silicon core spacing starting from ~980 nm. The MPSi array is assumed to be of cubic symmetry; hence, the unit cell area in this case will be about 1 μm^2 (see Figure 5.1a). The waveguide mode area will be 0.16 μm^2 for the 1550 nm wavelength. Hence, the coupling losses at the first MPSi layer interface (if pores and hence silicon cores are constant across all MPSi depth) will be around 84%. Such a value is not suitable for most of optical applications. Even modification of the porosity of the MPSi layer near both interfaces will not boost the potential transmittance to more than 25 to 35%. Hence, a better way of suppressing the cross-coupling while keeping coupling losses at a reasonable level is needed. This is possible by making further modifications in MPSi layer structure.

One relatively simple way to modify the MPSi layer structure to suppress cross-coupling while keeping the potential transmittance at a reasonable level is to fill the pores with highly reflective materials, such as metals. A thin layer of low refractive index material (for example, thermally grown silicon dioxide) is still needed between the metal filling the pores, and the silicon core waveguides to suppress the propagation losses in the waveguides. Such a design will also suppress light propagation through the MPSi structure other than through waveguides.

Let's again consider the MPSi array of cubic symmetry and near-square macropores as in Figure 5.1a. Before investigating the particular values of coupling efficiencies and cross-coupling coefficients, we need to define the sizes of silicon cores and the thickness of the silicon dioxide layer needed to maintain single-mode operations of silicon core waveguides and the metal-caused propagation losses on a sufficiently low level.

The optimal silicon core cross-section can be estimated from the previously given consideration (see Figures 5.3 and 5.4). Hence, we can set the maximum silicon core cross-section to be around 300 nanometers to ensure single-mode waveguiding at 1550 nm. However, we need to pay additional attention to propagation losses, absent in previous case but considerable here. As an example, in Figure 5.5 the calculated dependences of loss coefficients of TE-polarized waveguide modes on the silicon core cross-section are presented. The MPSi layer structure was assumed to be of cubic symmetry; the silicon dioxide layer thickness was assumed to be 200 nm, and calculations were made for 1550 nm wavelength. In the calculations the metal was assumed to be nickel.

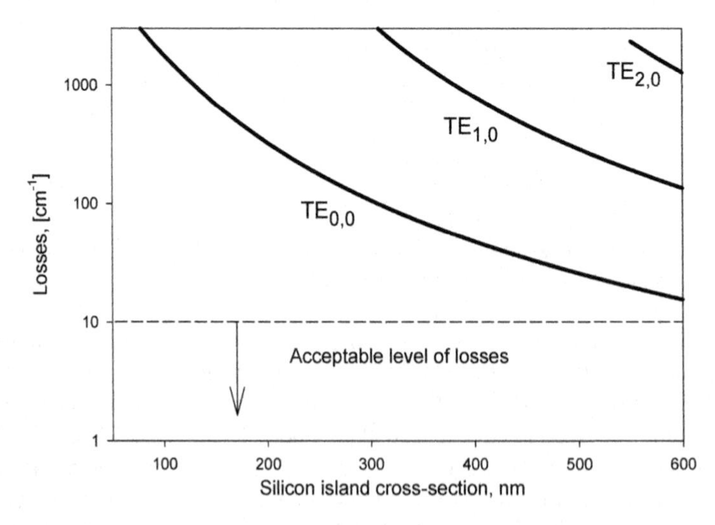

Figure 5.5. Calculated dependences of loss coefficients of TE polarization waveguide modes on the silicon core cross-section. After [2]

The waveguide mode losses decrease for each mode with the increase of the silicon core cross-section. It happens due to the increase of the localization of the waveguide mode in the silicon core and the consequent decrease of the localization of the waveguide mode near the metal/silicon dioxide boundary as soon as the waveguide parameters are getting further from the cut-off conditions. The important question here is: what level of losses is acceptable for providing a reasonable performance of the optical components, in this example for the 1550 nm band, based on such an MPSi layer. From one point of view, the coupling losses at the first MPSi layer interface will be dominating (not less than 20 to 30%); in comparison even high propagation losses of 10% will not cause a dramatic change of the performance of the MPSi-based optical component. Taking into account that the MPSi layer thickness will not exceed 300–400 μm, it will give maximum acceptable propagation losses of around 10 cm^{-1}. From another point of view, the quality of the Bragg resonance in the waveguide is a function of the level of losses in waveguide: high losses are particularly undesired for optical components that require sharp spectral features.

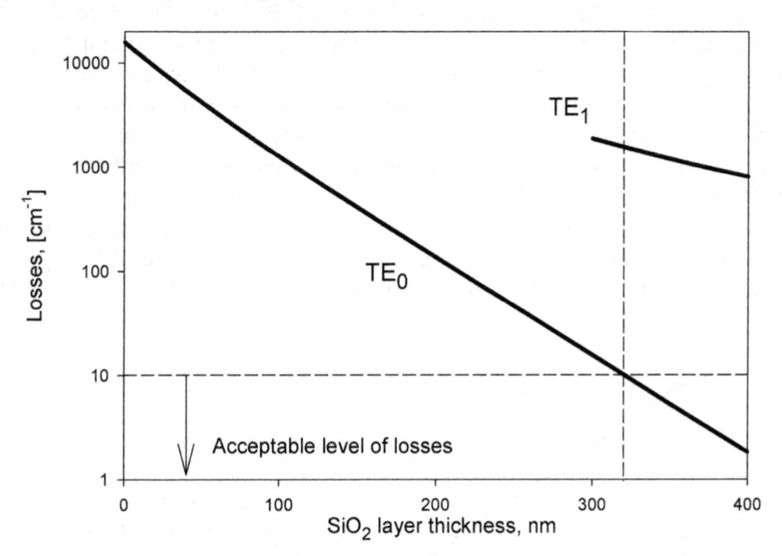

Figure 5.6. Calculated dependences of loss coefficients of TE polarization waveguide modes on the silicon dioxide layer thickness. After [2]

For the MPSi structure such as analyzed in Figure 5.5 the losses are higher than what is desired. However, the losses are known to drop exponentially with the "buffer layer" (silicon dioxide in this case) thickness. As an example, Figure 5.6 gives the calculated dependence of the loss coefficients of the TE polarization waveguide modes on the silicon dioxide layer thickness. The MPSi layer structure was assumed to be of cubic symmetry; the silicon core cross-section was assumed to be 275 nm, and calculations were made for 1550 nm wavelength. In the calculations the metal was assumed to be nickel. In this particular example the losses became acceptable for a silicon dioxide layer thick enough to cause the appearance of a second waveguide mode. However, unlike the MPSi layer with pores not filled with metal, it will not cause problems due to very high losses of this mode (two orders higher than that of fundamental mode). Hence, we can conclude that the optimal MPSi layer structure for optical component designed for 1550 nm spectral band should be as follows: the silicon core cross-section should be no more than 300 nm, and the silicon dioxide thickness should be no less than 300 nm.

With these parameters in mind, we are now ready to estimate the level of cross-coupling and get the approximate sizes of an MPSi array. In Figure 5.7 the calculated dependence of the cross-coupling coefficient on the silicon core separation is given for the fundamental waveguide mode. The wavelength was assumed to be 1550 nm, the silicon core cross-section was assumed to be 275 nm, the silicon dioxide layer thickness was assumed to be 300 nm, and the MPSi layer structure was assumed to be the same as in Figure 5.1a but having pores filled with nickel. By comparing Figures 5.4 and 5.7 one can see that metal in the pores indeed suppresses the cross-coupling between neighboring silicon core waveguides. While cross-coupling suppression is not very strong, it should provide better potential coupling efficiency at the first MPSi layer interface due to a smaller acceptable value of porosity at the same or better cross-coupling. We can estimate that it will be comparabel to the air-filled MPSi layer case. According to Figure 5.7, the minimum acceptable (from the viewpoint of the cross-coupling suppression) silicon core spacing of 910 nm corresponds to the area of such an MPSi array unit cell of ~ 0.82 μm^2. The waveguide mode area will be 0.2 μm^2 for the 1550 nm wavelength. Hence, the coupling losses at the first MPSi layer interface, if pores and hence silicon cores have constant sizes across all MPSi depth, will be ~ 75 %. With the modification of the porosity of the MPSi layer near both interfaces the potential transmittance can be increased to about 40 % to 50 %.

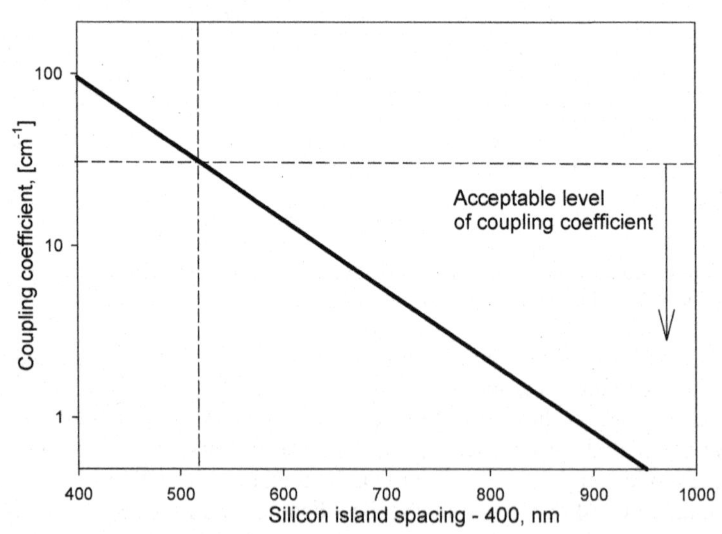

Figure 5.7. Calculated dependences of loss coefficients of cross-coupling coefficients of fundamental waveguide modes on the silicon cores separation for metal-filled pores. After [2]

This value of potential transmittance is already acceptable for some applications, but it is still lower than what is needed for, e.g., telecommunication applications. The appearance of metal-caused propagation losses in this design is negative from the viewpoint of the quality of the Bragg resonance. However, since the value of propagation losses is very different for different orders of waveguide modes, it is possible to utilize two-mode waveguides in such an MPSi layer, since a second-order mode will dissipate due to high losses. It provides an opportunity to utilize higher than 300 nm (for 1550 nm wavelength) cross-sections of silicon core waveguides for the same area of an MPSi array unit cell (i.e., it increases the coupling efficiency at the first interface of MPSi layer).

The calculated spectral dependences of the coupling, outcoupling, and propagation losses of a nickel-filled cubic MPSi array with a pore-to-pore distance of 1500 nm, pore diameters (after oxide layer growth step) of 750 nm, and a 300 nm oxide layer covering the pore walls, is presented in Figure 5.8. One can see that propagation losses are expected to be considerably smaller than coupling losses for such a material.

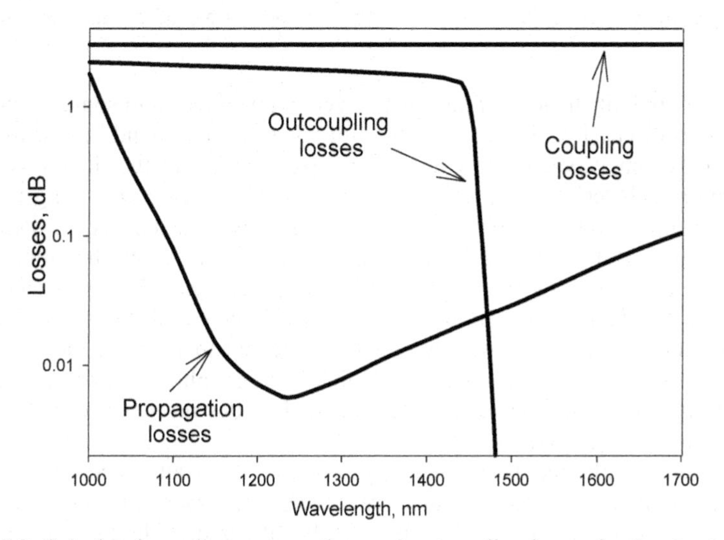

Figure 5.8. Calculated coupling, propagation, and outcoupling losses for the structure with parameters given in the text. After [2]

A possible benficial tradeoff between cross-coupling and coupling losses is to create MPSi arrays of more complex symmetries. Examples of such MPSi arrays are given in Figure 5.9. For the MPSi layers in Figure 5.9 the ratio of silicon core waveguide size to the MPSi array unit cell is considerably higher than that of the MPSi layers of Figure 5.1. It should be noted that macropores in those advanced-symmetry MPSi structures could be also filled either with air or with metal (if a stronger suppression of cross-coupling is required).

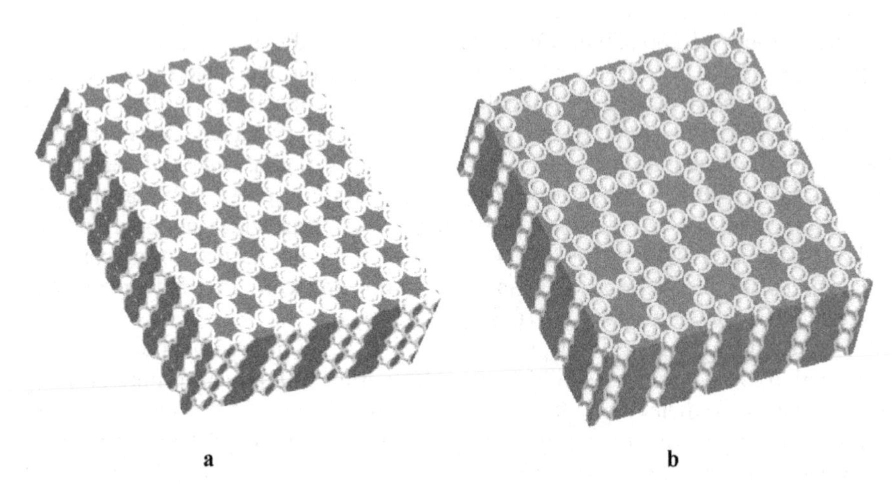

Figure 5.9. Possible symmetries of an MPSi array: **a** advanced hexagonal symmetry; **b** advanced cubic symmetry. In both cases the macropores are of circular cross-section. After [7]

The required dimensions of the MPSi layer can be inferred to some extent from the previous discussion. The cross-section of silicon cores should be less than 300 nm for 1550 nm wavelength. The size of macropores and the thickness of the silicon dioxide layer covering the pore walls should be found according to cross-coupling suppression requirement. In Figure 5.10 the calculated dependences of cross-coupling coefficients of fundamental waveguide modes on the silicon cores separation are given for the MPSi layer of Figure 5.9a. The wavelength was assumed to be 1550 nm, the pores were assumed to be circular, and the thickness of the silicon dioxide layer was assumed to be equal to halve of the macropore diameter. As follows from Figure 5.10 the cross-coupling coefficient between neighboring silicon core waveguide fundamental modes reaches acceptable levels at the silicon core separation of 790 nm.

By taking simple geometrical calculations and taking into account the single-mode requirement, we can find that the needed macropore diameter should be around 180 nm and the needed thickness of the silicon dioxide layer should be at least 120 nm. The unit cell area for an MPSi layer with such dimensions will be 0.26 μm^2, while the silicon core waveguide mode cross-section will be around 0.16 μm^2 to 0.2 μm^2. Hence the coupling losses at the first MPSi layer interface will not exceed 30%. It is expected that by reducing the porosity of the MPSi layer (i.e., by reducing the macropore diameters) near both interfaces of the MPSi layer it is possible to increase the potential transmittance of such a structure over 75%. As discussed for the simple symmetries of MPSi arrays previously, the macropores in the MPSi arrays of Figure 5.9 can be filled by metal. In this case the cross-coupling between neighboring silicon core waveguides is expected to be even smaller (i.e., the ratio of waveguide mode area to unit cell area can be increased). For such an array the potential transmittance is expected to exceed 85%, which is suitable for most of optical applications.

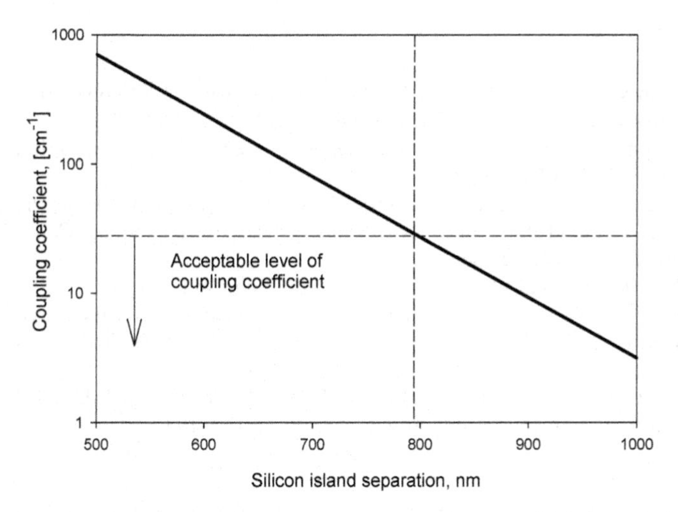

Figure 5.10. Calculated dependences of cross-coupling coefficients of fundamental waveguide modes on the silicon core separation for the MPSi layer of Figure 5.9a. After [2]

It should be noted that in all discussions in this section the coupling efficiency at the first interface of the MPSi layer was calculated as a ratio of the waveguide mode area to the MPSi array unit cell area. However, the reflectance at the air/silicon boundary (as well as the reflectance at the silicon/air boundary during outcoupling at the second MPSi layer interface) strongly degrades the coupling (and outcoupling) efficiency. Due to the high refractive index of silicon, reflection losses are expected to reach 31% at each MPSi layer interface at normal incidence. It means that even for optimized MPSi array symmetries the transmittance through such an MPSi array would not exceed 30% to 40%. Fortunately, the high reflection loss problem can be solved by coating both surfaces of the MPSi layer with a conventional antireflection layer. This could be a single-layer antireflection coating, like silicon monoxide, or a multilayer antireflection coating.

5.4 Conclusions

The recently suggested waveguide mechanisms of transmission through MPSi arrays can enable a number of potential optical applications of macroporous silicon in the IR range, some of which will be reviewed in more detail in Chapter 11. This mechanism can be expected in other porous semiconductors as well, e.g. in porous InP and GaAs [8]. Porous alumina layers can be expected to provide similar transmission channels even in visible wavelength range. While so far the only suggested application for such a transmission channel was optical filter application [2,7], one can expect that novel applications of this unique mode of transmission through porous semiconductors will be found in the future.

5.5 References

[1] Lehmann V, Stengl R, Reisinger H, Detemple R, Theiss W, (2001) Optical shortpass filters based on macroporous silicon. Appl. Phys. Lett. 78:589–591.

[2] Kochergin V, (2003) Omnidirectional Optical Filters. Kluwer Academic Publishers, Boston, ISBN 1-4020-7386-0.

[3] Marcuse D, (1974) Theory of Dielectric Optical Waveguides, Academic Press.

[4] Yariv A, Yeh P, (1984) Optical Waves in Crystals: Propagation and Control of Laser Radiation. John Wiley & Sons.

[5] Yeh P, (1988) Optical Waves in Layered Media. John Wiley & Sons.

[6] Okamoto K, (2000) Fundamentals of Optical Waveguides, Academic Press.

[7] Kochergin V, Föll H, (2006) Novel optical elements made from porous silicon. Review Materials Science and Engineering R, 52:93–140.

[8] Föll H, Langa S, Carstensen J, Christophersen M, Tiginyanu IM, (2003) Review: Pores in III-V Semiconductors. Adv. Materials, 15:183–198.

6

Optical Components from Mesoporous Silicon

6.1 Introduction

Porous silicon with pore sizes below the wavelength of the light in the micro-, meso, or lower macropore region (in what follows always addressed as "mesoporous") offers the opportunity to "engineer" the refractive index at the visible and the IR spectral range by variations of the porosity of the layer. This property can be utilized in a number of optical components. Note that the luminescence properties of microporous silicon [1] will not be considered here.

Reflective type of optical filters, or mirrors based on mesoporous silicon superlattices were proposed first by G. Vincent over a decade ago [2]. In this paper porous silicon super-lattices (roughly two microns thick) were etched on p-doped (100) oriented (8–15) Ωcm wafer; the porous layer was microporous silicon. The porosity and the concomitant refractive index modulations were formed by modulating the applied current density during the electrochemical etching. Independently, a similar approach was demonstrated by another group at the same year [3,4]. In [3] it was shown that in addition to modulating the current density porous silicon superlattices could also be formed by anodization of substrates with layers of different doping concentrations or different compositions at constant current density. Since then a large number of papers was published with respect to (meso)porous silicon filters and mirrors. Porous silicon filters were proposed to be used in color-sensitive photodiodes [5], luminescent devices [4], sensors [6–11]. A detailed review of these activities can be found in [12].

In this chapter we will review filter applications of mesoporous silicon. Particularly, we will focus on mesoporous silicon filters for the mid to far infrared region (light with wavelengths above 3 μm), where the advantages of the mesoporous silicon technology are most promising. We will start with brief description of optical filters in general, will briefly review far IR mesoporous filters and will then address one of the most challenging problems for this technology – environmental instability problem and the ways to mitigate this problem.

6.2 Interference Optical Filters

A wide variety of interference filters is used in different applications due to the possibility of engineering the optical properties (for example transmission and/or reflection spectra) of the filters by changing the structure of the multilayers comprising interference filter. We will review a few types of filters according to the utility function.

6.2.1 Antireflection Coatings

The simplest type of interference filter is an antireflection coating. An antireflection coating is usually understood to be a coating of the optical surface that minimizes the reflection from the surface at some angles of incidence over a specified wavelength range. The simplest type of antireflection coating is a *single-layer antireflection coating* or an *antireflection layer* (ARL). It can be shown that the reflectivity from the substrate/antireflection layer/superstrate system reaches a minimum when optical thickness D of the antireflection layer is equal to

$$D = n_1 d = \frac{m\lambda}{4\cos\theta_1},\ m = 1,3,5,... \tag{6.1}$$

where λ is the wavelength of light, θ_1 is the angle of incidence, n_1 is the refractive index of antireflective layer and d is the physical thickness of the antireflective layer. At the conditions of (6.1) and normal incidence the reflectivity becomes

$$R_{02} = \left(\frac{n_0 n_2 - n_1^2}{n_0 n_2 + n_1^2} \right)^2 \tag{6.2}$$

where n_0 is the refractive index of the incident medum and n_2 is the refractive index of the substrate.

Expressions (6.1) and (6.2) define both the optimal refractive index n_1 of the antireflective layer and its optimal thickness, which for the normal incidence takes the following values:

$$\begin{cases} n_1 = \sqrt{n_0 n_2} \\ d_1 = \dfrac{m\lambda}{4n_1}, m = 1,3,5,... \end{cases} \tag{6.3}$$

Under the conditions (6.3), the reflectance from the substrate becomes zero at normal incidence with the assumption of an absorption-less antireflection layer.

The most frequently used case is when the incident medium is air with the index of refraction being very close to unity. In this case n_l should be equal to

$(n_2)^{1/2}$ to get the optimal ARL performance. Typically the material with the value of refractive index closest to $(n_2)^{1/2}$ is used for the ARL.

As an example, let us consider an ARL for a silicon substrate (with refractive index around 3.5) at a wavelength of 1550 nm. According to Equation 6.3, the ARL refractive index should be around $(3.5)^{1/2} = 1.871$. Silicon monoxide (nominally SiO) is the best choice for an ARL since it has a refractive index that is close to the optimum ($n_{SiO} = 1.9$) and exhibits very low losses in the near and mid IR [13]. The antireflection properties of silicon monoxide and silicon dioxide layers are illustrated in Figure 6.1. The decrease in reflectivity is significant for the ARL from SiO_2, even though the refractive index of silicon dioxide ($n = 1.46$) is 23% smaller than the desired value of 1.871. The reflectivity of the silicon monoxide-coated silicon wafer with the right thickness of silicon monoxide can be considered to be zero for most applications.

To our belief it is improbable that porous semiconductors can find applications as single-layer antireflection coatings even so their index of refraction can be made just right, since, as illustrated in Figure 6.1, even significant deviations in refractive index of a conventional ARL from the optimum still provides good suppression of the reflection (at lower costs). However, we should note that porous silicon antireflection coatings were suggested for a number of applications with major emphasis placed on the reduction of reflection from solar cells (see, for example, references [14] and [15]).

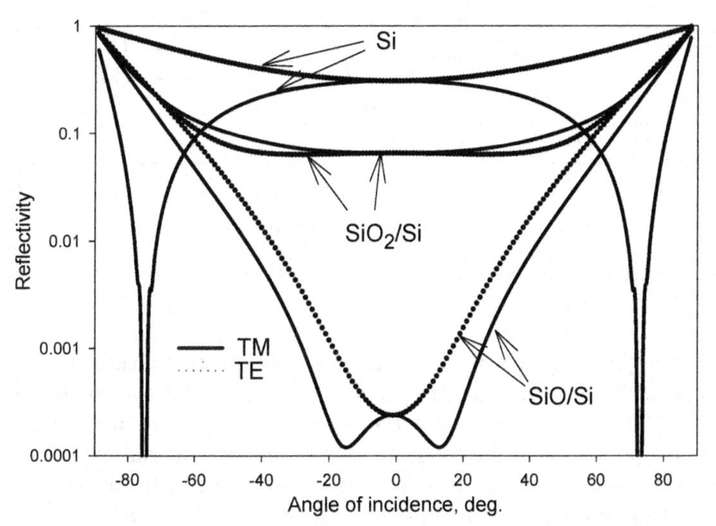

Figure 6.1. Angular reflectivity spectra for silicon, silicon coated by 204 nm of silicon monoxide ($n = 1.9$), and silicon coated by 269 nm of silicon dioxide ($n = 2.46$) at 1550 nm wavelength. After [16]

A typical disadvantage of single-layer antireflection coatings is the limited angular and spectral range over which the reflection is sufficiently suppressed, as illustrated in Figures 6.1 and 6.2. Also, the ARL performance at nonzero angles of incidence exhibits relatively strong polarization dependence. A number of applications (especially those dealing with wide spectral ranges) thus cannot be satisfied with single-layer antireflection coatings. For such applications *multilayer antireflection coatings* are used [17].

There is a wide variety of multilayer antireflection coating designs [17]. They differ by the number of layers used, by particular requirements for the reflection wavelength, angular spectra, and so on. The number of layers used varies from just two layers to eight or more layers, depending of the width of the spectral and angular bands where the reflection must be suppressed. The most frequently used multilayer antireflection coatings consist of alternative layers of high refractive index (in future consideration denoted as *H*) and low refractive index (denoted as *L*) materials. The antireflection coating structure can be written in schematic form, for example, {Air | *L H L H* | Glass} or {Air | *L H L H L* | Glass}.

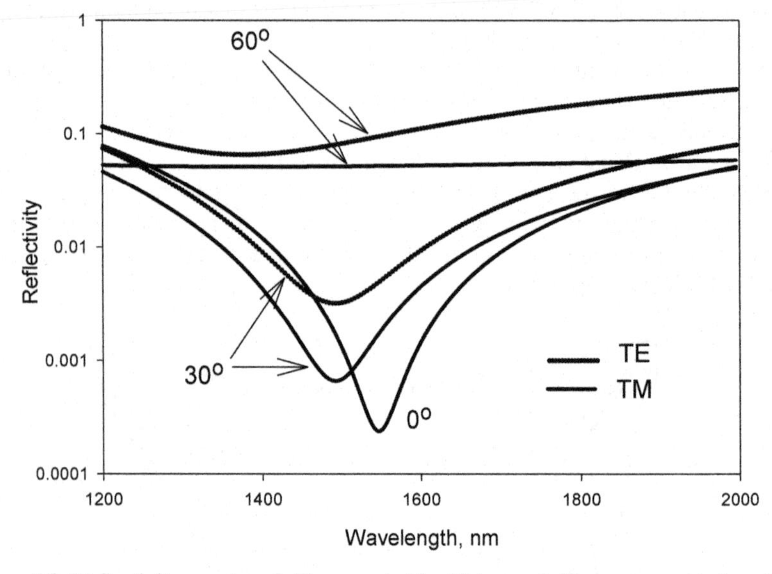

Figure 6.2. Reflectivity spectra of silicon coated by 204 nm of silicon monoxide (n = 1.9) at different angles of incidence. After [16]

As an example, let us consider a five-layer antireflection coating on glass (with refractive index of 1.52) that is designed to have a center wavelength around 600 nm. The low refractive index material in the antireflective multilayer will be magnesium fluoride, and the high refractive index material will be titanium dioxide. Figure 6.3 gives the calculated reflectivity spectra for different angles of incidence. The multilayer structure was taken from [17]: $n_0 = 1$; $n_1 = 1.38$ (magnesium fluoride); $d_1 = 126.643$ nm; $n_2 = 2.30$ (titanium dioxide); $d_2 = 41.377$ nm; $n_3 = 1.38$; $d_3 = 47.483$ nm; $n_4 = 2.30$; $d_4 = 27.230$ nm; $n_5 = 1.38$; $d_5 = 246.926$ nm; $n_6 = 1.52$. The suppression of the reflectance is not only very strong at the central wavelength of such an antireflection structure but also significant (below 1%) at a wide wavelength range – from about 500 nm to more than 800 nm. That is, the width of reflection valley exceeds 50% of the central wavelength, more than twice of what is achievable with a single-layer ARL. The angular range of the suppressed reflection is also very wide.

The refractive index engineering possible with porous semiconductors as discussed in Chapter 3 together with the complexity of the fabrication process for high-performance multilayer antireflection coatings may open up some applications for porous semiconductor technology in this respect. This was indeed already suggested for solar cell applications [18,19].

6.2.2 Multilayer Dielectric Reflectors

Another important application of interference structures are optical mirrors. The simplest, oldest and most frequently used type of such mirrors is the metallic mirror. The main drawback of metallic mirrors is small but finite absorption of light in a metal, unacceptable for high power laser applications. For these applications different types of optical mirrors are usually used – in particular mirrors based on *multilayer dielectric coatings*.

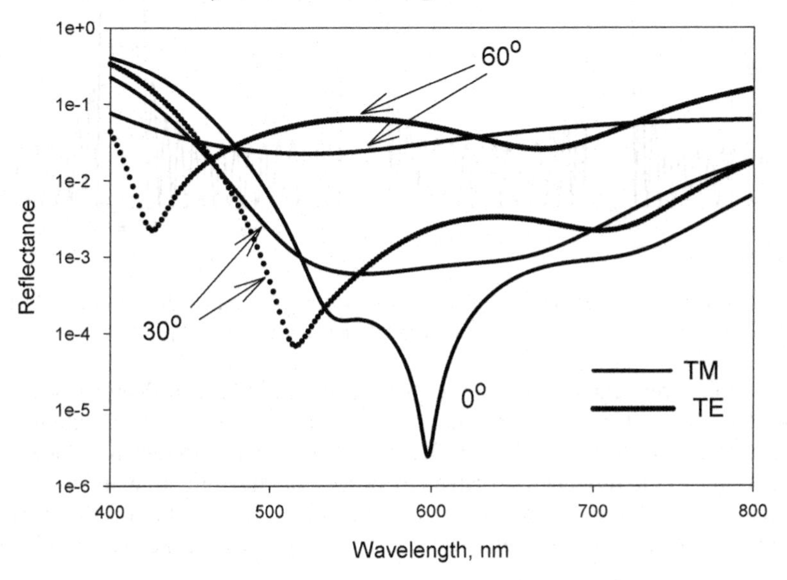

Figure 6.3. Reflectivity spectra of glass coated by five-layer antireflection coating at different angles of incidence. After [16]

Multilayer high-reflectance dielectric coatings are based on multiple-beam interference in a multilayer stack. The optical principle that serves as the base for multilayer dielectric reflectors is the multiple beam interference in multilayer dielectric structure.

The typical design of high-reflectance coatings is based on alternating quarter-wave layers of two different materials. The high reflectance in a quarter-wave stack takes place because the beams, reflected from all the interfaces in the multilayer, are in phase when they reach the front surface where the constructive interference of all the reflected waves occurs. For any number of layers the reflectivity is equal to [20]

$$R=\left(\frac{1-\dfrac{n_s}{n_0}\left(\dfrac{n_2}{n_1}\right)^{2N}}{1+\dfrac{n_s}{n_0}\left(\dfrac{n_2}{n_1}\right)^{2N}}\right)^2 \tag{6.4}$$

where n_1 and n_2 are the refractive indices of the alternative layers, n_0 is the refractive index of incident medium; n_s is the refractive index of the substrate, and N is the number of pairs of alternating layers in the stack. The reflectance according to (6.4) quickly approaches unity as the N grows.

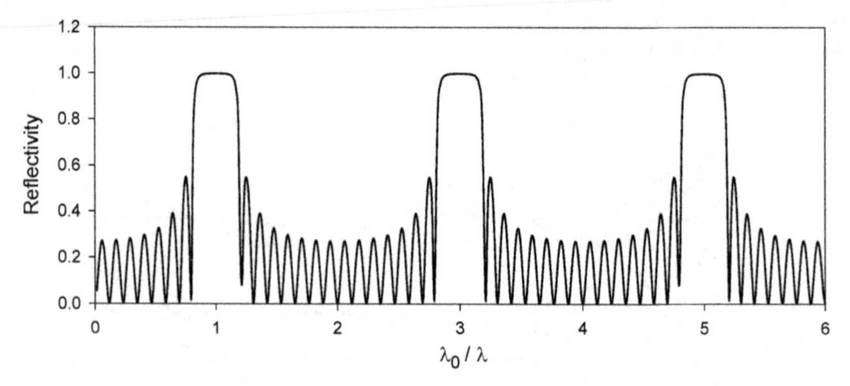

Figure 6.4. Calculated reflectance of a 17-layer stack of titanium dioxide (n = 2.3) and magnesium fluoride (n = 1.38) on glass at normal incidence. After [16]

The typical reflectance spectrum of a dielectric multilayer reflector is shown in Figure 6.4. The reflectance spectrum contains multiple peaks corresponding to the phase-matching conditions of the reflected waves at the different interfaces. The width of the high reflectance plateau depends on the refractive index contrast between low and high refractive index materials. The width of the first (or fundamental) high reflectance zone for quarter wave dielectric multilayer is [17]:

$$\frac{\Delta\lambda}{\lambda}=\frac{4}{\pi}\sin^{-1}\frac{n_H-n_L}{n_H+n_L} \tag{6.5}$$

where n_H and n_L are the refractive indices of high and low refractive index materials composing the high reflectance multilayer.

Extending the width of the reflection zone can be done by modifying the multilayer structure: One of these methods involves changing the thickness of successive layers throughout the multilayer to form a regular progression to ensure that at any wavelength within a reasonably wide range, there are enough layers in the multilayer that have an optical thickness sufficiently near to a quarter-wave to give high reflectance [21]. Another method involves the placement of a quarter-wave stack for one wavelength on top of another one for a different wavelength ([22]). To avoid Fabry-Perot type peaks in the resultant reflectance spectrum, it was found that a low index layer one quarter-wave thick at a mean wavelength should be placed in between stacks.

As an example, an SEM image of a mesoporous silicon multilayer reflector is provided in Figure 6.5a, while the photo of the mesoporous silicon multilayer reflector is shown in Figure 6.5b. The reflection spectrum from an 61-layer mesoporous silicon reflector is given in Figure 6.6. Apparent reflection in excess of 100 % is due to insufficiently accurate spectrometer calibration. The wide band ($\Delta\lambda/\lambda \sim 0.5$) of the high reflection region (> 90 %) is clearly demonstrated.

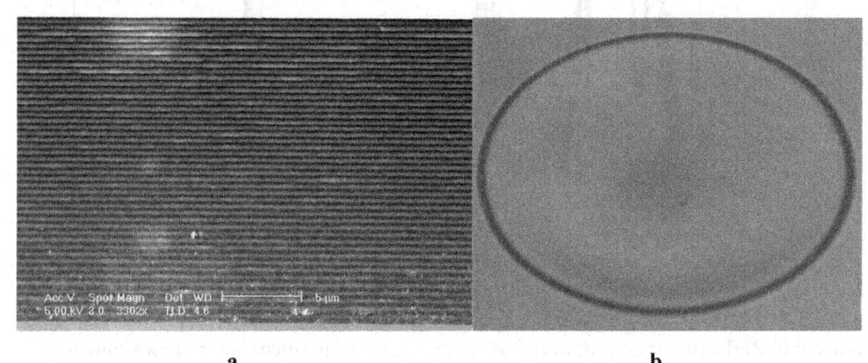

Figure 6.5. **a** SEM image of mesoporous silicon Bragg reflector (dark colour corresponds to high porosity low refractive index layer while light colour corresponds to low porosity high refractive index layers); **b** photo of 50mm diameter mesoporous Bragg reflector

Bragg reflectors and mirrors were among the first optical components demonstrated with porous silicon technology [2–4] and were significantly advanced over the years [23–26]. For example porous silicon reflectors and output couplers were demonstrated for continuous-wave and mode-locked Ti:Sapphire lasers for near IR wavelengths [27] with laser parameters achieved similar to those obtained with commercial mirrors. Dye lasers with porous silicon mirrors were also demonstrated [27].

Porous silicon with laterally varying reflection wavelengths (by manipulating the porosity and thickness of the silicon in the lateral direction) were recently demonstrated as well [28].

In addition to passive dielectric mirrors made from porous silicon, active (wavelength-tunable) porous silicon mirrors were also demonstrated by a number of groups. For example, tunable porous silicon mirrors were demonstrated by infiltrating porous silicon microcavities with liquid crystals [29–30].

Active MOEMS porous silicon membranes (freestanding porous silicon multilayers suspended on a MEMS structure with a thermal bimorph actuator that generates drive forces to move porous silicon layers) for beam steering, filtering, and splitting were demonstrated as well [31].

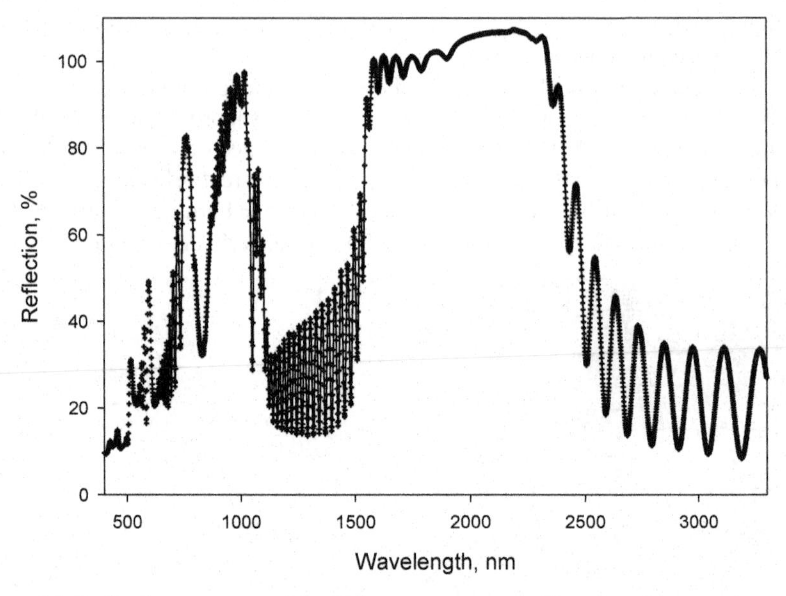

Figure 6.6. Reflection spectrum of the 61-layer mesoporous silicon interference mirror

Despite of potential advantages of porous silicon dielectric mirrors, such as refractive index engineering, good control over thicknesses of individual layers, potentially inexpensive fabrication of a large number of layers, and compatibility with silicon processing (i.e., potentially straightforward integration with CMOS technology), to the best of authors knowledge, none of such products made it to market as of yet. We believe that two main issues hamper the production of this technology to date: 1) environmental instability of mesoporous silicon (that will be addressed in more detail later in this chapter) and 2) relative inexpensiveness of already available multilayer dielectric reflectors for visible and near IR range. This statement does not apply, however, for mid and far IR filters that will be described in more detail layer in this chapter as well. While mesoporous silicon reflectors can be potentially cheaper than those fabricated with sputtering techniques, the cost difference is insufficient to offset the cost of development and commercialization. It thus still needs to be seen whether mesoporous silicon reflectors will make real commercial applications in the future.

6.2.3 Bandpass Filters

Another popular type of interference filters are band-pass filters. Such filters provide a wavelength-limited region of transmission surrounded on each side by rejection regions. Band-pass filters can be divided into *broadband-pass filters* and *narrowband-pass filters* according to the width of the transmission zone but there is no definite boundary between these two types of band-pass filters. In this book we follow Macleod [17] and assume that a broadband-pass filter has bandwidth of 20% or more of the central wavelength of transmission band. Broadband-pass filters are typically constructed by placing two multilayer Bragg reflector stacks (short-pass and long-pass) on the opposite sides of a single substrate.

Narrowband pass filters are typically constructed by incorporating a "cavity" or spacer layer inducing a phase shift into a quarter wave structure as illustrated in Figure 6.7. Two different cases of such a filter should be considered: {Air| *H L H L H L* ***H H*** *L H L H L H*| Substrate} and {Air| *H L H L L H* ***L L*** *H L L H L H*| Substrate} and the refractive indices of layers adjacent to air and substrate should be high to maximize the reflection from the multilayers. The transmission spectrum of such a structure will have a narrow peak within the broad valley. As an example, in Figure 6.8 the calculated transmittance spectrum of a 17-layer titanium dioxide/magnesium fluoride narrowband-pass filter on a glass substrate is presented for a high refractive index (titanium dioxide) spacer layer.

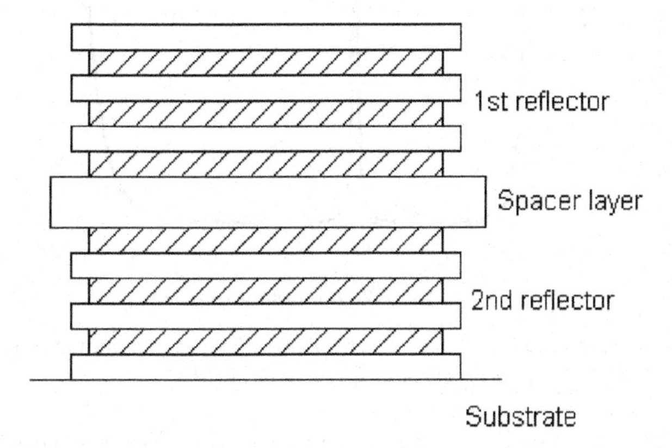

Figure 6.7. The structure of a single cavity all-dielectric narrow bandpass filter

The width of the peak and the transmittance at the peak depend on the reflectivities of the two multilayer stacks. As shown in [17] the bandwidth of a single-cavity interference narrowband pass filter for a high refractive index spacer layer is:

$$\frac{\Delta\lambda_H}{\lambda_0} = \frac{4n_L^{2x} n_s}{m\pi n_H^{2x+1}} \tag{6.6}$$

where n_L and n_H are the refractive indices of the low- and high-refractive index layers in the multilayer stacks; n_s is the refractive index of the spacer layer; λ_0 is the wavelength of the transmittance peak; x is the number of high-index layers in each stack (both stacks assumed to have equal number of layers); and m is the order of the stack. For a low-refractive index spacer layer, the bandwidth is:

$$\frac{\Delta\lambda_L}{\lambda_0} = \frac{4n_L^{2x-1} n_s}{m\pi n_H^{2x}} \tag{6.7}$$

As follows from (6.6) and (6.7) the bandwidth of the single-cavity interference narrowband pass filter decreases with the number of layers in the stacks and with the index contrast between the low- and high-refractive index layers in the stack.

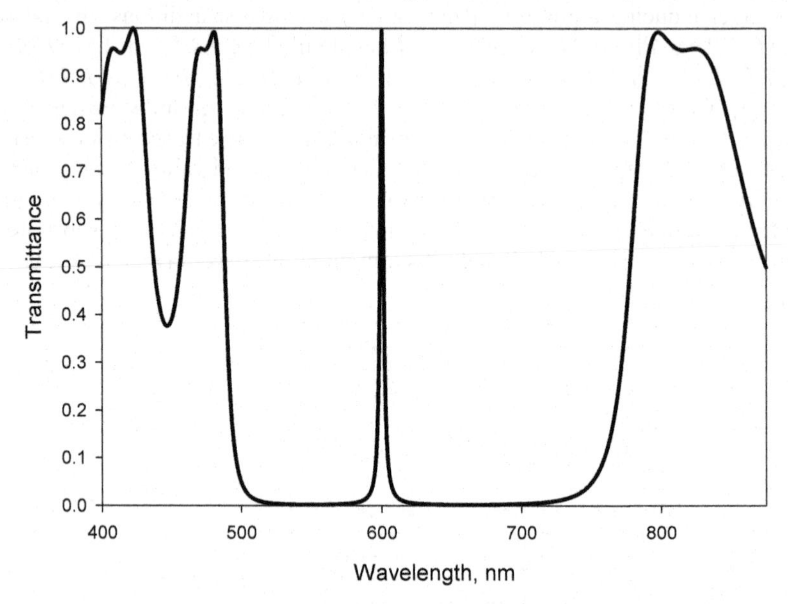

Figure 6.8. Calculated transmittance spectrum of 17-layer titanium dioxide/magnesium fluoride narrowband-pass filter on glass substrate. After [16]

As illustrated by Figure 6.9, the spectral shape of the transmission peak for single-cavity narrowband pass filter is not ideal. For filter purposes a nearly rectangular shape of transmittance spectra is desired. In addition, the maximum achievable rejection in the rejection zone of the filter and the bandwidth of the transmission zone are related. The solution of these problems was found in *multiple-cavities* filter designs.

The simplest type of multiple cavities filters is the double-cavity filter with the structure of {*Air* | *reflector* | *half-wave spacer* | *reflector* | *half-wave spacer* | *reflector* | *Substrate*}. An example of the calculated transmittance spectra of titanium dioxide/magnesium fluoride multilayer on the glass substrate narrowband-pass interference filters having one and two cavities is given in Figure 6.9.

The advantages of the double-cavity design with respect to the single-cavity design are obvious from Figure 6.9. To achieve even steeper edges and a flatter top of the transmission peak (to suppress the peaks at both sides of pass band, so-called *rabbit's ears*) filters with even larger number of cavities are used.

As an example of an optimized filter structure with strongly suppressed "rabbit ears", Figure 6.10 shows the calculated transmittance spectrum of a 70-layer seven-cavity narrowband-pass interference filter at normal incidence.

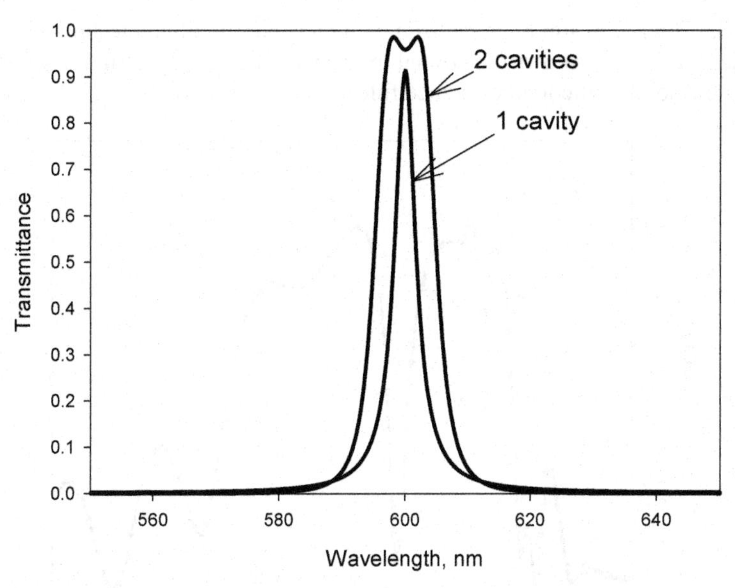

Figure 6.9. Calculated transmittance spectra of titanium dioxide/magnesium fluoride multilayer on the glass substrate narrowband-pass filters having one and two cavities "HH" (19 layers for single cavity and 25 layers for double cavity {*Air* | *H L H L H L HH L H L H L H L H L H L HH L H L H L H* | *Glass*}). After [16]

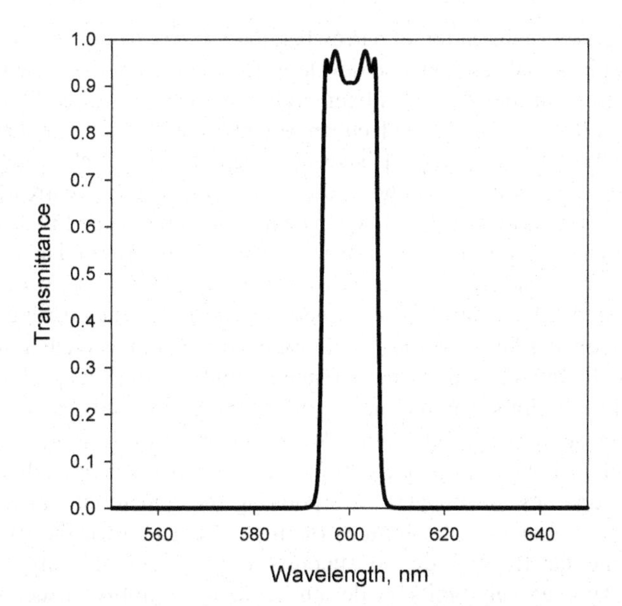

Figure 6.10. The calculated transmittance spectrum of a 70-layer seven-cavity narrowband-pass filter at normal incidence. After [16]

Similar filters can surely be realized with mesoporous silicon technology. As an example, Figure 6.11 shows the normal incidence reflection spectrum from a single cavity mesoporous silicon interference filter.

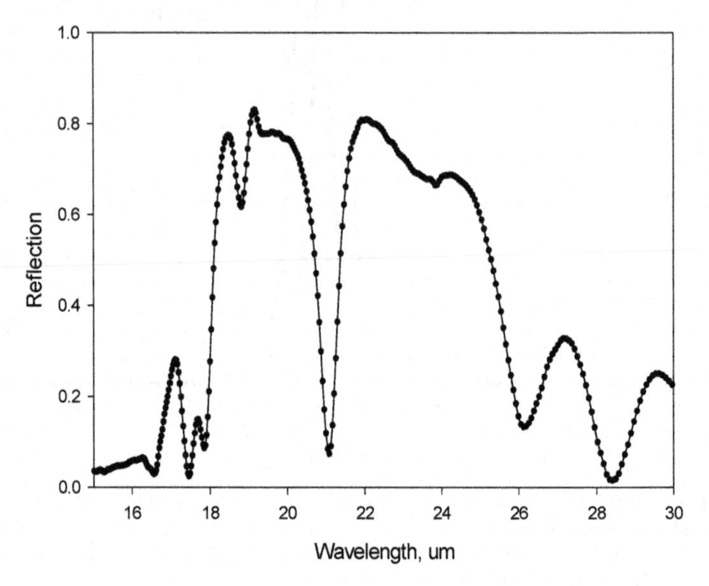

Figure 6.11. Reflection spectrum from mesoporous silicon narrowband pass filter

The simplicity of fabrication of narrowbandpass filters from mesoporous silicon resulted in a substantial research and development effort. Single-cavity interference filter structures from mesoporous silicon were demonstrated as early as 1994 [4]. The directionality, narrowing, and enhancement of both the photoluminescence [32] and electroluminescence [33] have been reported for such structures. Even subnanometer linewidths (around 900 nm wavelength) narrowband pass mesoporous silicon structures were experimentally demonstrated [34,35]. Multiple-cavity multilayer mesoporous silicon structures were demonstrated as well [35].

A potentially attractive application of mesopore technology to optical filters was suggested by the team of Lake Shore Cryotronics, Inc. (Columbus, Ohio, USA): interference filters for the far IR range (> 20 μm wavelengths) [36,37]. Available far IR filters based on interference in multilayer stacks presently serving this need have serious shortcomings. First, there are only a few materials sufficiently transparent beyond 15 μm. Moreover, most of these materials are not environmentally stable, have poor adhesion and quite different crystallographic and mechanical properties. Multilayers incorporating these materials are subjected to delamination at cryogenic temperatures (temperatures at which the filters have to be used in the far IR in order to suppress thermal emission from the filters themselves). This severely limits the design freedom and utility of such filters.

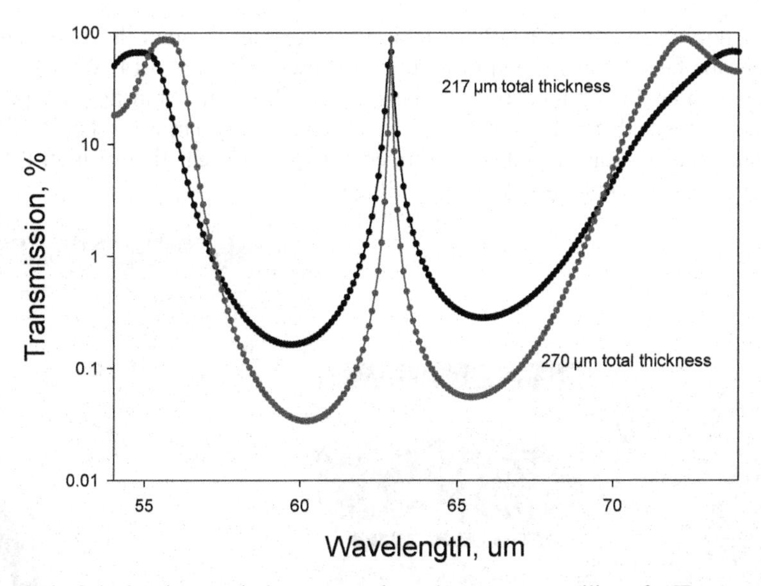

Figure 6.12. Calculated transmission spectra through mesoporous silicon far IR narrowband pass filters with different numbers of layers (29 and 35 for 217 µm thick and 270 µm thick samples respectively) in quarter wave stacks

A related problem that has not been overcome either is the roughness amplification as the thickness of the film is increased, which produces a columnar structure that is not optically stable. As a result, such filters tolerate poorly rapid temperature changes and high or low temperatures. The columnar structure also causes scattering of light and poor image transfer through the filters. Another consequence of the materials problems is that the maximum sizes of the mid-to-far IR multilayer interference filters are severely limited. The common maximum size is around 1 inch diameter.

Mesoporous silicon multilayer structures, on the other hand, offer a sufficient level of transparency down to 200 µm, excellent mechanical and thermal stability (due to the absence of mechanical stresses at layer interfaces) and the capability to make large-size filters in a cost-effective way. The filters do not exhibit delamination problems and are well suited for operation at extreme temperatures and temperature gradients. Because all the layers are made of single crystal silicon, they have similar mechanical and thermal properties (particularly thermal expansion coefficients) and there are thus no adhesion problems.

The calculated transmission through a specifically designed mesoporous silicon far IR narrowband pass filter is shown in Figure 6.12. It is apparent that a more than 200 µm thick mesoporous silicon multilayer is needed to provide viable far IR filters [36,37]. As an example, Figure 6.13 gives SEM images of the cross-sections of different mesoporous silicon far IR interference optical components. Light color corresponds to low porosity high refractive index layers while dark color corresponds to high porosity low refractive index layers. Careful optimization of etching conditions indeed resulted in good mechanical quality mesoporous silicon multilayers with a thickness up to 380 µm.

The normal incidence FTIR transmission spectrum through the free-standing mesoporous far IR narrowband pass filter is shown in Figure 6.14. While there is definitely room for optimization of the rejection level, the transmission peak is clearly visible and can be considered to be sufficiently large for far IR filters. Environmental testing of such filters indeed prove the prediction of high thermal stability of such an optical components [36,37].

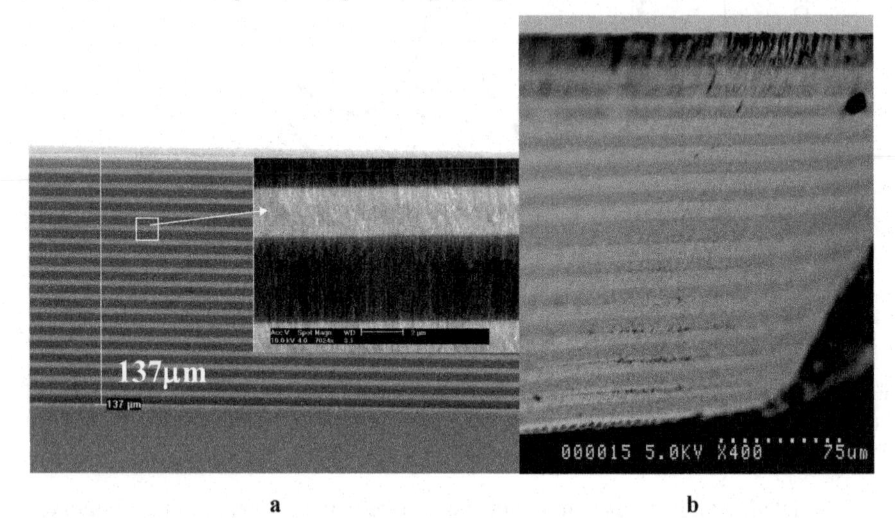

Figure 6.13. SEM images of cleaved mesoporous silicon far IR multilayer dielectric reflector (**a**) and narrowband pass filter (**b**)

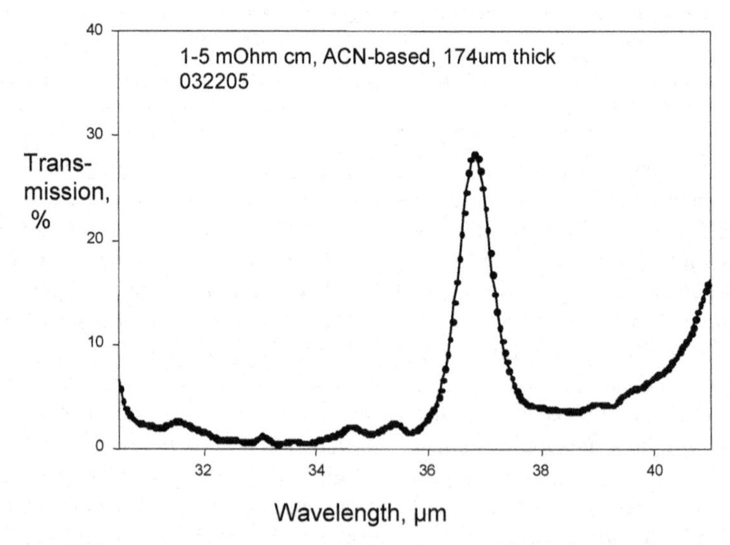

Figure 6.14. FTIR transmission spectrum of mesoporous silicon far IR narrowband pass filter

While significant progress was made in the development of mesoporous silicon interference filters, two major problems hampered the commercialization of mesoporous bandpass filters to date: 1) environmental instability problems, and 2) substrate absorption problems. We will review the state of the art in solving these problems below.

6.3 Environmental Instability Problem

Environmental instability of mesoporous silicon originates from the very high surface area of mesoporous silicon (600–700 m^2/cm^3 for microporous and 200–250 m^2/cm^3 for mesoporous silicon [38,39]. It causes the following unwanted effects over time:

1. Blue-shift of the filter wavelength.
2. Degradation of the filter performance (quarter wavelength matching of the layers gets destroyed).
3. Decrease of reflectivity.

In infrared transmission-type filters a strong degradation of the transmission is also observed. In general, two different causes are responsible for the effects listed above.

The blue-shift of the filter wavelength and degradation of the quarter-wave matching (causing some decrease of the reflectivity as well) is tied to the relatively slow formation of native silicon dioxide on the pore walls [40]. It is well known that at room temperature and ambient environment about 7 nm of native oxide can be formed with time. Due to the very large surface area of mesoporous silicon, the slow formation of native oxide on all the pore walls is causing very strong optical effects.

The loss of transparency of the mesoporous silicon in the mid and far IR range (it is only transparent through all mid and far IR ranges immediately after etching) is associated with the adsorption of polar gas molecules on the freshly etched silicon surface [41], as illustrated schematically in Figure 6.16. This effect is much faster than the oxide formation effect and a substantial loss of the transparency in mid and far IR can be observed within a couple of hours after the drying of the sample. In about one week this effect reaches saturation. However, this effect is also fully reversible within some time after etching (for at least two weeks) as illustrated in Figure 6.15.

During the last decade several approaches were suggested for solving the environmental instability problem. This includes the pre-oxidation of mesoporous silicon sample either chemically (by boiling in HNO_3) or thermally (see, e.g. [40]. The reported results suggested that none of these techniques succeeded in a complete prevention of the aging effect. For thermal oxidation a set of conditions exists (at quite elevated temperatures) at which the process is slowed down to acceptable levels for most applications. It should be noted, however, that such a processing results in very substantial shifts and reshaping of the reflection and transmission characteristics of the mesoporous silicon multilayer. Also, due to a substantial lattice mismatch of silicon dioxide and silicon, thermally oxidized layers (especially thick ones, such as required for mid and far IR filters) exhibit large stresses even at room temperatures (causing considerable bowing of the wafer) and the effect is even stronger at cryogenic temperatures.

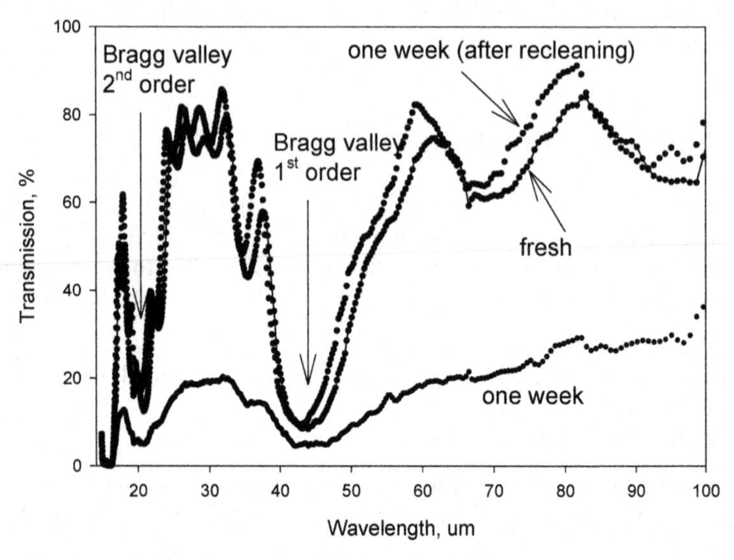

Figure 6.15. Transmission spectra through a mesoporous silicon multilayer membrane taken after etching, after 1 week aging in ambient atmosphere and after recovering the membrane in wet chemical solution. After [42]

Considerable efforts were devoted towards stabilizing the porous silicon with organic compounds. The basis of these efforts is the replacement of the relatively unstable hydrogen termination of the Si surfaces of pore walls (Si-H bonds) by more stable carbon termination (Si-C bonds). A quite thorough review can be found in [43]. Various techniques were employed for these purposes, including hydrosilylation [44–46], electrochemical grafting [47] and a method based on alkyl halides [48] to name a few. These methods mainly were devoted toward the stabilization of porous silicon optical properties in the visible spectral range. In the far infrared spectral range the advantages of these methods need still to be proven, since FTIR studies show the appearance of additional absorption bands associated with Si-C bonds and (sometimes) other bonds related to the particular composition of pore wall coverage.

Another possible solution for the stabilization of the mesoporous silicon multilayer is the use of a sintering process. Sintering of the mesoporous silicon is usually performed in oxygen and water vapor-free hydrogen or hydrogen/inert gas atmosphere at elevated temperatures (up to 1100 °C and above) and results in a complete closure of the pores at the top surface of the mesoporous silicon layer and sometimes at the porosity interfaces (see, e.g. [49–52]). Such a technique can accomplish a substantial stabilization of the mesoporous silicon multilayer, indeed, while offering a much-reduced mechanical stress level compared to thermally oxidized wafers. This is no surprise, of course, since the mesoporous silicon layers are still all-silicon after the sintering process. However, the reshaping/shifting of the spectral characteristics of the samples were still considerable and quite comparable to thermally oxidized samples.

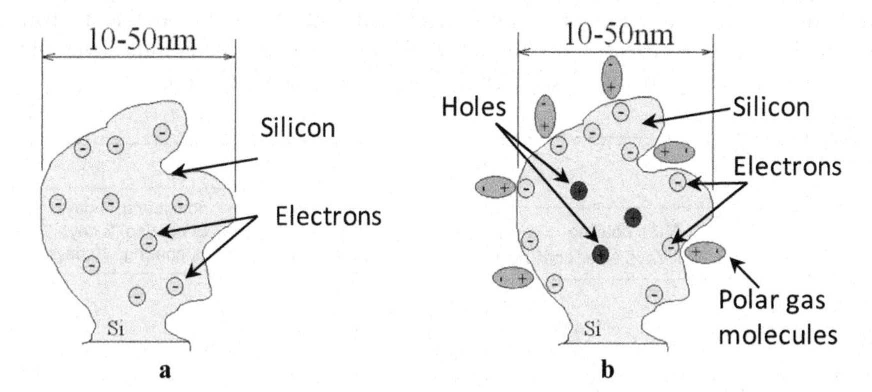

Figure 6.16. Schematic drawings illustrating the increase of the absorption in mesoporous silicon due to reversible adhesion of polar gas molecules, attracting electrons, thus creating uncompensated charge carriers (holes) in the mesoporous silicon. After [42]

Figure 6.17. SEM image of a silicon encapsulating layer sputtered onto the mesoporous silicon surface. After [42]

Coating the mesoporous silicon surface with layers of SiO_2 deposited at relatively low temperatures by Plasma Enhanced Chemical Vapor Deposition (PECVD) or by Low Pressure Chemical Vapor Deposition (LPCVD) were suggested in [53]. In the same reference, spin-coated PMMA layers were also suggested for the protection of mesoporous silicon mirrors from moisture and it was observed that PMMA is actually better suited for moisture protection, while PECVD deposited SiO_2 is better suited for the coating of mesoporous silicon membranes.

In [36,37,42] it was proposed to use magnetron-sputtered amorphous silicon (a-Si) layer on the mesoporous silicon surface as a method of improving the environmental stability of the far IR filters. The advantage of such an approach is based on the fact that the thermal expansion coefficient of a sputtered silicon layer will be the same, or at least close to that, of the mesoporous silicon multilayer and silicon substrate. In other words, the filter structure will be all-silicon.

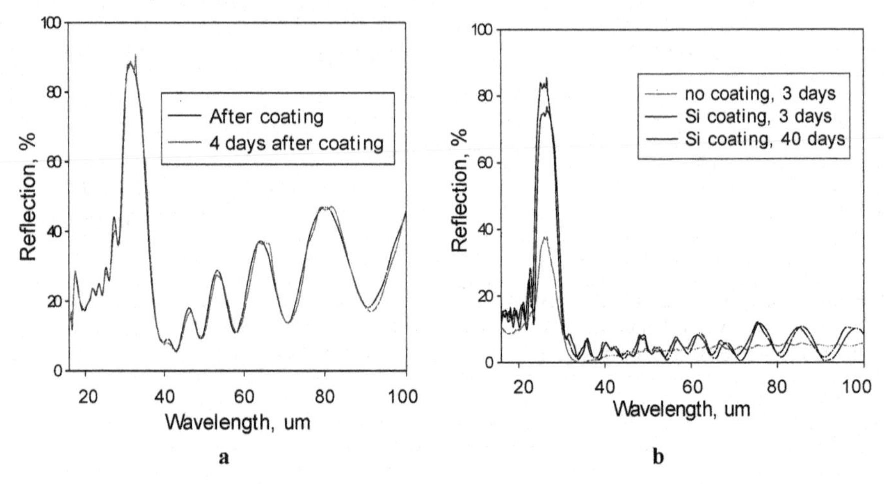

Figure 6.18. a Reflection spectra of a mesoporous silicon filter with the surface protected by 200 nm a-Si taken at different times after the deposition. **b** Reflection spectra of a mesoporous silicon filter with and without a 200 nm a-Si layer on the mesoporous silicon surface taken under three days aging conditions and under forty days aging conditions. After [42]

Here we show the results of this method. The change of the mesoporous silicon filter characteristics due to the DC magnetron sputtered a-Si protective coating is minor, as illustrated in Figure 6.19. A SEM image of such a filter structure is given in Figure 6.17. The good quality of the silicon layer on the top of the mesoporous silicon multilayer is apparent. Optical testing of the filters confirmed that for a sputtered silicon layer with a thickness of 200 nm or larger, the environmental stability of the filter improves considerably. This statement is illustrated by Figure 6.18. As follows from Figure 6.18a, no change in the reflection from a mesoporous silicon quarter wave stack is detectable three days after the deposition, while the reflectivity of the uncoated sample at the Bragg wavelength degrades strongly as shown in Figure 6.18b. However, forty days aging in ambient atmosphere resulted in some degradation of the reflectance, indicating that some further optimization of the encapsulation is still required.

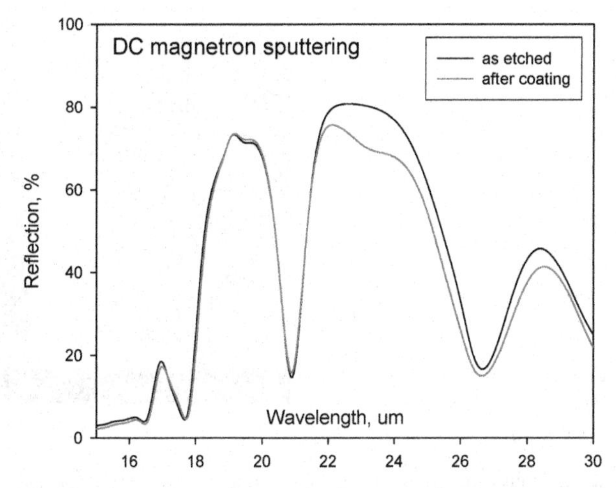

Figure 6.19. FTIR reflection spectra from a far IR narrowband pass mesoporous silicon filter before and after the DC magnetron sputtering of Si protective coating

6.4 Substrate Absorption Problem

A big problem with mesoporous silicon IR filters in a transmission mode comes from the free carrier absorption (also known as multiphoton absorption) in the silicon substrate remaining underneath the mesoporous layer. Highly doped silicon wafers with resistivities in the range of 1 to 20 mΩcm are totally opaque in most of mid IR and far IR spectral range at room temperature. While in [54] it was suggested that the transparency of the un-etched substrate underneath the porous layer improves at low temperatures; even at cryogenic temperatures the transparency of the substrate cannot be restored in the mid IR part of the spectrum. The solution for this problem for the last ten years has been the removal of the mesoporous silicon multilayer from the unetched silicon substrate by means of a current spike at the end of the etching process, which form a cavity and thus "lifts off" the mesoporous silicon filter structure. Such a free-standing mesoporous silicon multilayer is termed a "membrane". Although some mesoporous silicon membranes are relatively robust and flexible (if sufficiently thick, as for far IR filters), most will not survive this procedure i. Thinner or higher porosity contrast membranes are usually unacceptable brittle for practical applications.

In [36], a new approach was suggested. Instead of a silicon wafer with the same doping all the way through, it was decided to use a two layer silicon wafer composed of a highly doped Si layer, with the proper thickness for the mesoporous layers, bonded to low-doped "handle" wafer. This provides the high carrier concentration necessary for mesoporous multilayer formation in the high doped layer, which, after porosification, is backed by silicon that is highly transparent throughout the far IR range. Thus, with such an approach, the free carrier absorption problem in the nonporous silicon disappears and the mesoporous silicon will be mechanically reinforced by highly transparent, high resistivity silicon. A schematic drawing of such a bonded structure is shown in Figure 6.20a.

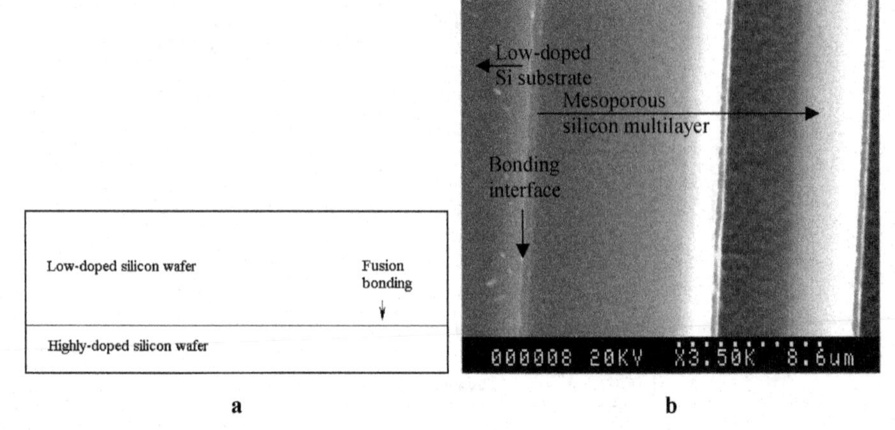

Figure 6.20. **a** Schematic drawing of the fusion bonded silicon wafer used for fabricating transmission-type mesoporous silicon IR filters; **b** SEM image of a bonded wafer cross-section after electrochemical etching up to the bonding interface. The bonding interface is indicated. After [42]

The key point in choosing a wafer bonding technique suitable for these purposes is the necessity for an uniform and generally highly conductive bonding interface. A fusion bonding technique is capable of meeting this requirement. In fusion bonding, the two polished sides of two silicon wafers of proper doping, thickness and orientation are brought into contact in a clean environment, followed by some annealing under some pressure to increase the bond strength [55]. The absence of native oxide on the wafer surfaces is a must to insure good conductivity across the bonding interface.

The backside of the bonded wafer combination was metalized and the wafer front side electrochemically etched. The etching was stopped exactly at the bonding interface, which was made possible by monitoring the current/voltage characteristics of the system. An SEM image of a portion of the mesoporous silicon filter around the bonding interface is given in Figure 6.20b. In order to prove the transparency of the mesoporous silicon multilayer with a handle wafer, one of the etched wafers was examined in an FTIR at room temperature. The transmission spectrum is shown in Figure 6.21.

One can see that the structure is indeed transparent in the far IR range. The lower than expected level of transmission within the narrow transmission band is actually caused by losses in the mesoporous silicon multilayer itself, originating from the environmental instability of the mesoporous silicon as described above since the FTIR measurements were performed some time after etching.

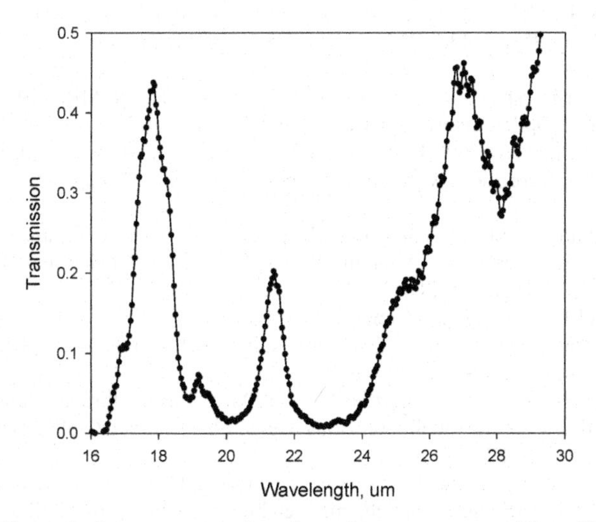

Figure 6.21. Transmission spectrum taken through a mesoporous silicon filter etched on a bonded wafer at angle of incidence ~ 14° from normal at room temperature. After [42]

6.5 Conclusions

Mesoporous silicon filters offer a number of important advantages compared to filters fabricated with conventional deposition techniques. A number of problems encountered with these filters slowed down the market introduction of such filters. However, it appears that some recently proposed solutions to these problems are capable to resolve these issues, and mesoporous silicon filters are finally expected to enter the commercial market in the very near future.

6.6 References

[1] Canham LT, (1990) Silicon quantum wire array fabrication by electrochemical and chemical dissolution of wafers. Appl. Phys. Lett. 57: 1046–1048.

[2] Vincent G, (1994) Optical properties of porous silicon superlattices. Appl. Phys. Lett. 64:2367–2379.

[3] Berger MG, Dieker C, Thoenissen M, Vescan L, Lueth H, Muender H, Theiss W, Wernke M, Grosse P, (1994) Porosity superlattices: A new class of Si heterostructures. J. Phys. D. 27: 1333–1336.

[4] Frohnhoff S, Berger MG, (1994) Porous silicon superlattices. Adv. Mat. 6: 963–965.

[5] Krueger M, Berger MG, Marso M, Reetz W, Eickhoff T, Loo R, Vescan L, Thönissen M, Lüth H, Arens-Fischer R, Hilbrich S, Theiss W, (1997) Color-sensitive Si-photodiode using porous silicon interference filters. Jpn. J. Appl. Phys. 36: L24.

[6] Sailor MJ, (1997) Sensor application of porous silicon, in: L.T. Canham (Ed.), Properties of Porous Silicon, IEE-Books, London.

[7] Lauerhaas JM, Credo GM, Heinrich JL, Sailor MJ, (1992) Reversible Luminescence Quenching of Porous Si by Solvents. J. Am. Chem. Soc. 114:1911–1912.

[8] Sailor MJ, Credo G, Heinrich J, Lauerhaas JM, (1994) Method for Detection of Chemicals by Reversible Quenching of Silicon Photoluminescence. U.S. Patent 5,338,415.
[9] Lin VS-Y, Motesharei K, Dancil KS, Sailor MJ, Ghadiri MR, (1997) A Porous Silicon Based Optical Interferometric Biosensor. Science 278:840–843.
[10] Cunin F, Schmedake TA, Link JR, Li YY, Koh J, Bhatia SN, Sailor MJ, (2002) Biomolecular screening with encoded porous-silicon photonic crystals. Nat.Mat. 1:39–41.
[11] Li YY, Cunin F, Link JR, Gao T, Betts RE, Reiver SH, Chin V, Bhatia SN, Sailor MJ, (2003) Polymer Replicas of Photonic Porous Silicon for Sensing and Drug Delivery Applications. Science 299:2045–2047.
[12] Pavesi L, (1997) Riv. Nuovo Cimento 20:1–76.
[13] Cox JT, Haas G, (1958) Antireflection Coatings for Germanium and Silicon in the Infrared *J., Opt. Soc. Am.*, **48:**677–678.
[14] Matic Z, Bilyalov RR, Poortmans J, (2000) Firing through Porous Silicon Antireflection Coating for Silicon Solar Cells. physica status solidi (RRL) – Rapid Research Letters, 182:457–460.
[15] Lipinski M, Bastide S, Panek P, Lévy-Clément C, (2003) Porous silicon antireflection coating by electrochemical and chemical etching for silicon solar cell manufacturing. physica status solidi (a) 197:512–517.
[16] Kochergin V, (2003) Omnidirectional Optical Filters. Kluwer Academic Publishers, Boston, ISBN 1-4020-7386-0.
[17] Macleod H.A., *Thin-Film Optical Filters*, 3rd ed., Institute of Physics Publishing, 2001.
[18] Striemer CC, Fauchet PM, (2002), Dynamic Etching Of Silicon For Broadband Antireflection Applications. Appl. Phys. Lett. 81:2980–2982.
[19] Aroutiounian VM, Martirosyan KS, Hovhannisyan AS, Soukiassian PG, (2008), Use of porous silicon for double- and triple-layer antireflection coatings in silicon photovoltaic converters. J. of Contemporary Physics C, 43:72–76.
[20] Yeh P., *Optical Waves in Layered Media*, John Wiley & Sons, 1988.
[21] Heavens OS, Liddel HM, (1966) Staggered Broad-Band Reflecting Multilayers. *Appl. Opt.*, **5:**373–376.
[22] Turner AF, Baumeister PW, (1966) Multilayer Mirrors with High Reflectance Over an Extended Spectral Region. *Appl. Opt.*, **5:**69–76.
[23] Chan S, Fauchet PM (1999), Silicon Interference Filters And Bragg Reflectors For Active And Passive Integrated Optoelectronic Components. Proc. SPIE, 3630:144–154.
[24] Chan S, Tsybeskov L, Fauchet PM, (1999) Porous Silicon Multilayer Mirrors And Microcavity Resonators For Optoelectronic Applications. Mat. Res. Soc. Symp. Proc. 536, 117–122.
[25] Agarwal V, del Rio JA, (2003), Tailoring the photonic band gap of a porous silicon dielectric mirror. Appl. Phys. Lett. 82:1512–1514.
[26] Lugo JE, Lopez HA, Chan S, Fauchet PM, (2002), Porous silicon multilayer structures: a photonic band gap analysis. J. Appl. Phys. 91:4966–4972.
[27] Zheng WH, Reece P, Sun BQ, Gal M, (2004) Broadband laser mirrors made from porous silicon. Appl. Phys. Lett. 84:3519–3521.
[28] Hunkel D, Butz R, Ares-Fisher R, Marso M, H. Lüth H, (1998), Interference filters from porous silicon with laterally varying wavelength of reflection. J. of Luminescence, 80:133–136.
[29] Weiss SM, Haurylau M, Fauchet PM (2003), Tunable Porous Silicon Mirrors for Optoelectronic Applications. Mat. Res. Soc. Symp. Proc. **737:**529–534.

[30] Weiss SM, Fauchet PM (2003), Electrically Tunable Porous Silicon Active Mirrors. Phys. Stat. Sol. (a) **197:**556–560.
[31] Song D, Tokranova N, Gracias A, Castracane J, (2008) New approaches for chip-to-chip interconnects: integrating porous silicon with MOEMS. J. Micro/Nanolith. MEMS MOEMS 7:021013.
[32] Pavesi L, Mazzoleni C, Tredicucci A, Pellegrini V, (1994), Controlled photon emission in porous silicon microcavities Appl. Phys. Lett. 67:3280–3282.
[33] Chan S, Fauchet PM, (2001) Silicon microcavity light emitting devices. Opt. Mater. 17:31–34.
[34] Reece PJ, Lerondel G, Zheng WH, Gal M, (2002) Optical microcavities with subnanometer linewidths based on porous silicon. Appl. Phys. Lett. 81:4895–4897.
[35] Ghulinyan M, Oton CJ, Bonetti G, Gaburro Z, Pavesi L, (2003) Free-standing porous silicon single and multiple optical cavities. J. of Appl Phys. 93:9724–9729.
[36] Kochergin V, Sanghavi M, Swinehart PR, (2005), Porous silicon filters for low-temperature far IR applications. Proc. SPIE 5883:184–191.
[37] McGovern WR, Kochergin V, Sanghavi MR, Swinehart PR, (2006) Porous silicon optical filters for far-IR and low-temperature applications, Proc. Great Lakes Photonics Symposium.
[38] Sugiyama H, Nittono O, (1990) Microstructure and lattice distortion of anodized porous silicon layers. J. Crystal Growth 103:156–163.
[39] Bomchil G, Halimaoui A, Herino R, (1988) Porous silicon: the material and its application to SOI technologies. Microelectronic Eng. 8:293–310.
[40] Krüger M, Hilbrich S, Thönissen M, Scheyen D, Theiß W, Lüth H, (1998) Suppression of ageing effects in porous silicon interference filters. Optics Communications 146:309–315.
[41] Timoshenko VY, Dittrich Th, Lysenko V, Lisachenko MG, Koch F, (2001) Free charge carriers in mesoporous silicon. Phys. Rev. B 64:085314.
[42] Kochergin V, Föll H, (2006) Novel optical elements made from porous silicon. Review Materials Science and Engineering R, 52:93–140.
[43] Buriak JM, (2002) Organometallic Chemistry on Silicon and Germanium Surfaces. Chemical Reviews 102:1271–1308.
[44] Buriak JM, Allen MJ, (1999) Photoluminescence of Porous Silicon Surfaces Stabilized Through Lewis Acid Mediated Hydrosilylation. J. Lumin. 80:29–35.
[45] Buriak JM, Stewart MP, Allen MJ, (1998) Hydrosilylation Reactions on Porous Silicon Surfaces. Mater. Res. Soc. Symp. Proc. 536:173–178.
[46] Canham LT, Reeves CL, Newey JP, Houlton MR, Cox TI, Buriak JM, Stewart MP (1999) Derivatized Mesoporous Silicon With Dramatically Improved Stability in Simulated Human Blood Plasma. Advanced Materials 11:1505–1509.
[47] Robins EG, Stewart MP, Buriak JM, (1999) Anodic and Cathodic Electrografting of Alkynes on Porous Silicon. J. Chem. Soc., Chem. Commun., 2479–2480.
[48] Lees IN, Lin H, Canaria CA, Gurtner C, Sailor MJ, Miskelly GM, (2003) Chemical Stability of Porous Silicon Surfaces Electrochemically Modified with Functional Alkyl Species. Langmuir, 19:9812–9817.
[49] Sato N, (2003) Semiconductor Substrate and method for producing the same. U.S. Patent 6,593,211.
[50] Sato N, (2002) Method and apparatus for etching a semiconductor article and method of preparing a semiconductor article by using the same. U.S. Patent 6,413,874.
[51] Müller G, Brendel R, (2000) Simulated annealing of porous silicon. phys. stat. sol. (a) 182:313–318.
[52] Müller G, Nerding N, Ott N, Strunk HP, Brendel R, (2003) Sintering of porous silicon. phys. stat. sol. (a) 197:83–87.

[53] Corban R, Bousack H, Bohn HG, (2003) Protective coatings for interference filters made of porous silicon. phys. stat. sol. (a) 197:370–373.
[54] Christophersen M, Kochergin V, Swinehart PR, (2004) Macroporous Silicon filters for mid-to-far IR range. Proc. SPIE 5524:158–168.
[55] Gösele U, Tong Q-Y, (1998), Semiconductor wafer bonding. Annual Review of Materials Science, 28:215–241.

7

Long Wave Pass Filters

7.1 Introduction

Random arrays of macropores in Si scatter light at wavelengths smaller than the average pore geometry, e.g., the pore-to-pore distance. This chapter describes how this property can be used for making an infrared long-wave pass filter (LWPF). Such a filter would scatter light with wavelengths below an edge wavelength that is defined by the porous layer geometry and morphology, but transmit the light effectively and uniformly for wavelengths above this edge.

Infrared LWPFs have a number of important applications. They are used both in combination with narrowband pass filters (in order to achieve a narrow pass band with wide and deep rejection bands) and by themselves in applications where the radiation from shorter wavelengths should be suppressed in order to improve the signal-to-noise performance of detectors (as in astronomy, Fourier transform spectroscopy, etc).

7.2 Common Types of Long Wave Pass Filters

There are two main types of edge filters (long wave pass or short wave pass) with two principles of operation: absorption-based filters (i.e., filters where the rejection of light is caused by absorption in the filter material) and interference-based filters (i.e., filters where the rejection of light is caused by reflectance from multiple layers composing the filter) [1].

Absorption filters consist of a thin film or slide of material that has an absorption edge at the required wavelength. The typical example of a long-pass absorption filter are filters that utilize semiconductor materials. Semiconductors are known to have an absorption band that extends to some characteristic wavelength, which corresponds to the bandgap energy of a particular semiconductor. Since the absorption band edge of different semiconductors and semiconductor composites can vary from ~ 500 nm for gallium phosphide (GaP) and aluminum arsenide (AlAs) to more than 2 microns for indium arsenide (InAs) and InSb, and, more important, the absorption band edge can be smoothly tuned by adjusting the semiconductor composition (for example, the $Al_xGa_{1-x}As$ absorption band edge tunes quite linearly from 2.1 eV for x = 1 to 1.4 eV for x = 0), long-pass filters can be obtained with a reasonably sharp edge for any wavelength in the ~500 to 2400 nm range.

Other materials that are used to form absorption filters include coloured-glass filters (for example, Schott glass filter), which operate through the process of either ionic absorption of inorganic material dispersed uniformly through the glass slide or through the absorptive scattering of crystallites formed within the glass. Such filters offer quite wide design freedom in terms of absorption edge position, and they can be engineered either in short-pass or long-pass forms (contrary to semiconductor based filters, which are essentially long-pass filters).

The common disadvantage of all absorption-based edge filters is their instability under high-power (and even medium-power) radiation at the wavelengths of the absorption band. For most laser applications such filters are not suitable. Another important disadvantage of absorption-based edge filters in applications beyond visible and near IR wavelengths is the discrete set of absorption and transmission bands, thus limiting severely application of absorption-type filters in mid and far IR. For such applications a different type of edge filters is usually used – interference edge filters.

"Quarter-wave" stacks, i.e. multilayer strucurs with layer thicknes ¼ of the relevant wavelength can be considered as the basic type of interference edge filters. The transmission spectrum of quarter-wave stack contains alternative low- and high-reflectance zones and therefore alternative high- and low-transmittance zones. Edge wavelengths can be tuned by changing the layer thickness,. Such an edge filter will be suitable for relatively narrow-band applications that is, when the width of the rejection zone is larger than the spectral width of light to be eliminated. For all other cases, required elimination of all wavelengths shorter than (or longer than) a particular value, a different filter design is needed. Such an edge filter can be constructed by coupling the interference filter with the previously described absorption filter.

By using combined interference/absorption filter design, one can obtain filters that have the deep rejection of absorption filters and the flexibility in edge position and sharpness of interference filters. This approach is widely used in filters for UV-mid IR wavelength range. However, in the far IR this approach is not as successful as in the visible range because the number of transparent materials that can be used for multilayer stacks is rather small, and because of the large thickness of the required multilayers.

For the far IR range a scattering-type filter was already proposed in [2], utilizing a layer of diamond particles (or other transparent materials) of suitable sizes, which are spread on the surface of a sheet or substrate of a material that is transparent in the desired IR region (e.g. polyethylene, quartz or sapphire). Light with wavelengths around or below the size of the particles in the "active" layer is blocked by means of scattering, while light with wavelengths above the size of the particles is transmitted. Such filters are truly long wave pass filters, i.e. the transmission above the rejection edge is uniform over the wavelength region where substrate and particles are transparent. Such filters can be used both in room temperature and cryogenic applications. However, the mechanical stability of such filters is usually poor, since the active layer can be easily peeled off or will flake off. Moreover, the versions made with thin polymer films cannot withstand large pressure differentials very well, thus requiring long pumping times . Another drawback with particle-based filters is that in order to fabricate filters with a specific rejection edge position, particles of specific size are needed, which are not always commercially available. Also, in order to efficiently scatter the light the particles should have a fairly high refractive index, which makes polystyrene particles (which are inexpensive and widely available) not suitable for such applications.

7.3 Long Wave Pass Filters Made of Macroporous Silicon

Long Wave Pass Filters made on the basis of porous silicon technology were recently suggested [3] and successfully commercialized by Lake Shore Cryotronics [4,5]. While such filters are based on similar optical effects as the diamond particle filters, MPSi LWP filters have shown a number of important advantages.

From the MPS point of view, silicon has a sufficiently high refractive index in the IR spectral range (around 3.5) and the pores, if etched on a flat, unstructured surface, form a random array with a relatively narrow distribution of pore sizes that can be easily fine-tuned in a wide range by choosing suitable substrate and etching parameters [6]. Hence, one can expect that such a material will exhibit long wave pass behaviour similar to diamond particle filters [2]. The principle of operation is schematically illustrated in Figure 7.1a: the light with wavelengths 3 is long enough compared to the average interpore distance and is transmitted through the filter 1 (with some reflection losses), since it "sees" the macroporous layer 2 as an effective transparent medium. The light 4 with wavelengths lower than the average interpore distance experiences strong scattering (5) and thus is effectively blocked.

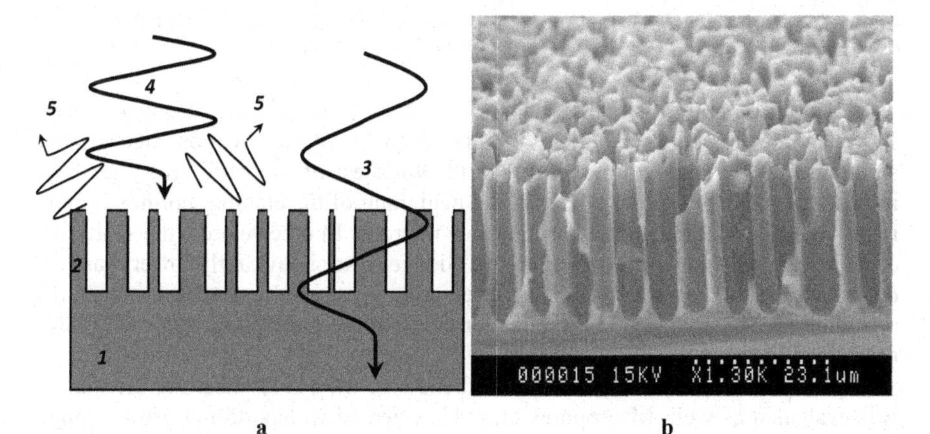

Figure 7.1. a Schematic drawing illustrating the principle of operation of a macroporous silicon LWPF. **b** SEM image of a suitable macroporous layer etched on (100) oriented silicon

Initial efforts were devoted toward the realization of a LWPF on p-doped (100)-oriented silicon wafers. An SEM image of a random MPSi array etching in organic electrolyte is presented in Figure 7.1b. An optical evaluation (see Figure 7.2) confirmed that a random MPSi arrays etched on (100) p-doped wafers exhibit well pronounced long wave pass behaviour, indeed. It was found that by changing the silicon doping, electrolyte concentration, current density and temperature during the electrochemical etching process, it is possible to fabricate good-quality (optically and mechanically) long wave pass filters with edge positions anywhere from ~3 μm to 35 μm.

The sharpness of the rejection edge could also to be controlled within wide limits by adjusting the total thickness of the MPSi layer and by optimizing (usually lowering) the temperature during electrochemical etching. Temperature control also allows to some extent to influence the spread of the average pore distance distribution in the array.

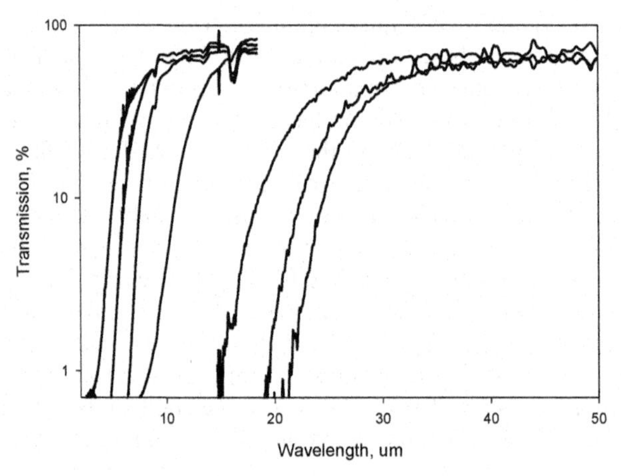

Figure 7.2. Normal incidence transmission spectra through a number of far IR LWPFs made from macroporous silicon layers etched on (100) oriented silicon wafers

It should be noted that MPSi layers etched on (100)-oriented p-doped silicon showed some drawbacks as well. While the sharpness of the rejection edge showed a pronounced increase with increasing of the thickness of the MPSi layer (i.e., with increase of the depth of the pores) in the beginning of the etching, not much more improvement is obtained after the first few microns. In other words, the scattering efficiency in the deeper portion of the MPSi layer was significantly lower than that on the MPSi layer/air interface. This observation provides the practical limit to the rejection edge sharpness and the rejection level of the LWPF fabricated on (100)-oriented silicon.

For a possible improvement of this behavior, the (111) substrate orientation was evaluated as well. Macropores on (111)-oriented wafers do not grow straight and perpendicular to the wafer surface, but rather in the three equivalent <113> directions [7]. They also show considerable branching, as illustrated by SEM images in Figure 7.3. Since the basis of MPSi LWPF is light scattering, such layers should be beneficial for LWPF applications. Unlike the case of MPSi layers etched on (100) oriented wafers, where the most efficient scattering takes place on the surface of the porous layer, in MPSi layers etched on (111) wafers one can expect high scattering efficiency of light throughout the thickness of the macroporous layer.

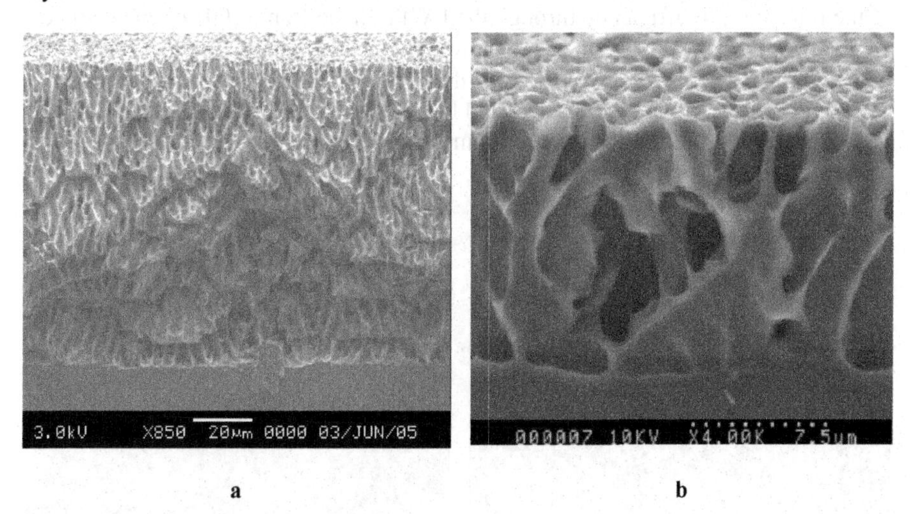

Figure 7.3. SEM cross-sectional views of a macroporous layer formed on p-doped (111)-oriented silicon wafers on different scales

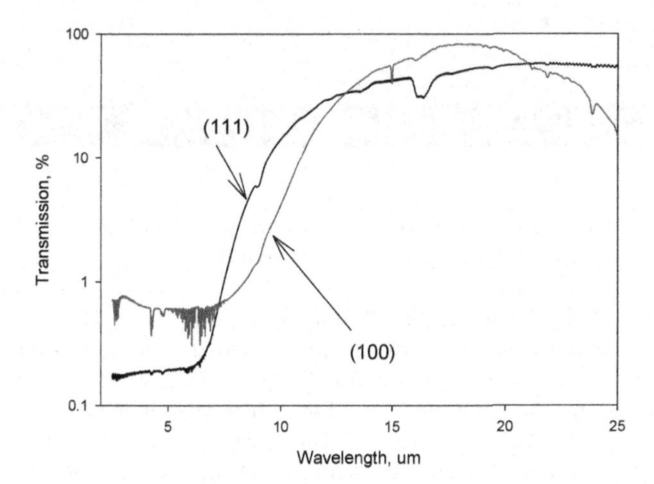

Figure 7.4. Normal incidence transmission spectra through far IR LWPFs made from MPL etched on (100) and (111)-oriented silicon wafers

For a comparison of the optical characteristics of MPSi layers etched on (100) and (111) oriented silicon wafers, macroporous layers with the same thickness were electrochemically etched on wafers of both orientations. Normal incidence transmission spectra of these samples, measured with a FTIR spectrometer, are shown in Figure 7.4. As expected, the branchy structure of MPSi layers on a (111) substrate caused much more effective scattering, indeed. The same thickness of an MPSi array provided stronger rejection and a sharper edge on a (111)-oriented wafer as compared to a (100)-oriented wafer. It should be noted that the reduction of the transmission efficiency through the LWPF made from (100)-oriented silicon at wavelengths above 20 μm is probably related to a measurement artifact. The higher transmission efficiency through the (100)-oriented MPSi layer above the scattering edge compared to that of (111)-oriented sample can be related to the higher porosity of the (100)-oriented sample and therefore lower reflection losses at the air/MPSi layer interface.

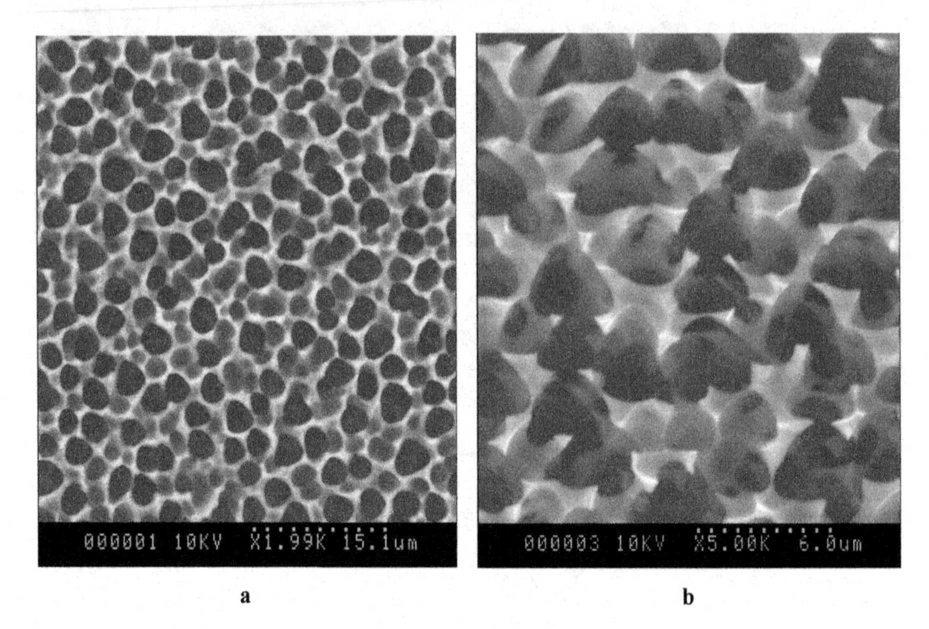

Figure 7.5. SEM images of two MPSi LWPF fiter surfaces. Both filters were fabricated on (111)-oriented p-doped wafers. **a** layers of medium porosity and **b** high porosity

The reflection losses on the air-MPSi interface in (111)-oriented wafers can be minimized by optimizing the electrochemical etching conditions similarly to the case of (100) oriented wafers described in Chapter 2. Experiments show that there is a certain limit for the initial porosity of the MPSi layer etched on a (111) wafer. Exceeding this limit leads t either to a significant reduction of the mechanical stability of the MPSi layer or even to etching it off completely. The variation of the MPSi surface structure with porosity is illustrated in Figure 7.5, which shows layers of medium porosity (Figure 7.5a) and high porosity (Figure 7.5b). It appears that the pore surface termination in the case of (111)-oriented wafers provides lower protection against the chemical dissolution as compared to the case of (111)-oriented wafers. Nevertheless, the much higher scattering efficiency of a MPSi layer etched on (111)-oriented wafers compared to that of (100)-oriented wafers provides the opportunity to use significantly thinner MPSi layers to achieve an acceptable sharpness of the rejection edge and rejection efficiency at short wavelengths. This in turn results in better mechanical and environmental stability of long wave pass filters made of MPSi layers etched on (111)-oriented wafers.

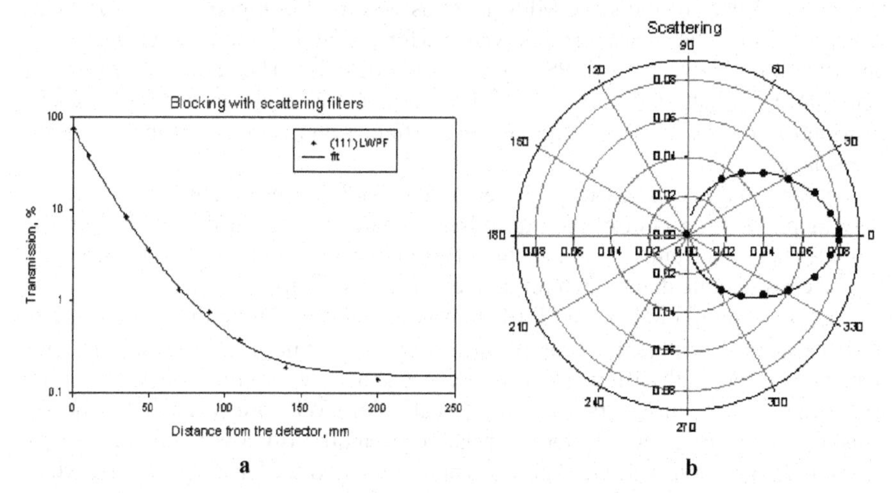

Figure 7.6. a Transmission through the MPSI LWPF within the rejection band as a function of detector-to-filter separation; **b** angular dependence of the scattering efficiency of the MPSi LWPF

It is important to note the difference in performance of scattering type Long Wave Pass Filters and absorption or interference filters. Transmission characteristics of the absorption or interference filters do not depend on the separation between the filter and detector surface (assuming no back-reflection from the detector surface is present) unless the detector is in the near field of the filter (with a separation in the range of just a few wavelengths). This is because light with wavelengths within the rejection band of such filters is either absorbed in the filter material (as in absorption filters) or reflected by the multilayer stack (as in interference filters). In scattering filters the light within the rejection band is scattered into some angular cone (defined by the morphology of scattering centers) and the rejection level depends on the intensity of the scattered light still captured by the photo detector surface.

It should be specifically noted that transmission spectra through different MPSi layers provided in this chapter were taken with an FTIR spectrometer at about 30 cm separation between the filter and the detector. The dependence of the rejection of the MPSi layer vs. MPSi layer-to-detector distance is shown in Figure 7.6a. The measurements were made with a 1550 nm Laser (far enough from the Si absorption edge but well within the rejection region of MPSi layer) at normal incidence. The detector had an area of ~2 mm × 2 mm. An MPSi layer etched on a (111)-oriented wafer with the rejection edge centered ~5 μm was used. This plot clearly demonstrates that in order to achieve significant rejection, one has to use at least some 10 cm of separation between the MPSi layer and the detector. This peculiar property of MPSi LWP filters should be carefully considered in applications of such filters.

Another important parameter of MPSi LWP filters is the angular distribution of the scattered light within the rejection band. The measured angular distribution of the scattered light of the same MPSi layer as was used in Figure 7.6a. is provided in Figure 7.6b. The measurements were performed at 1550 nm wavelength at 10 cm separation between the MPSi layer and the detector. The scattered intensity is very close to that of an "ideal" diffuser, characterized by an angular intensity distribution of cos(θ). This indicates very efficient and isotropic scattering in the MPSi layer.

An MPSi layer can be easily formed on both sides of the silicon wafer. This not only helps to better control the transmission characteristics of the filters, but also provides the opportunity to reduce the reflection losses within the pass band (which can be considerable due to high refractive index of silicon).

The pass band of MPSi LWPF filters was found to be flat up to a wavelength of 200 μm at room temperature, as illustrated by the normal incidence transmission spectra of filters with different rejection edges shown in Figure 7.7. The reason for this is that the silicon wafers used for fabrication of LWPF are typically low doped, thus the free-carrier absorption is relatively small. Moreover, the number of vibronic bands in silicon is relatively small compared to that of polymers, which have been used with LWPFs based on colloidal diamond [2].

In addition, LWPFs made from MPSi layers are mechanically stable and "all-silicon", and thus can be operated in a wide temperature range, including "bake-out" to cryogenic temperatures.

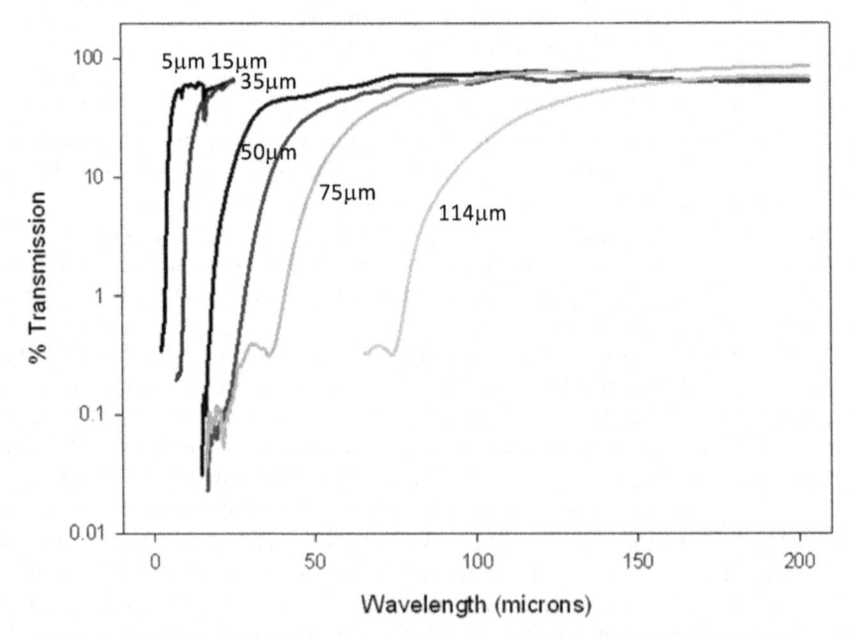

Figure 7.7. Normal incidence transmission spectra through far IR MPSi LWPFs

Figure 7.8. Photo of the far IR MPSi LWPF (at glancing angle), illustrating the good surface quality of macroporous silicon

7.4 Conclusions

Macroporous silicon long-wave pass infrared filters offer a number of important advantages compared to other kinds of long-wave pass IR filters. The advantages are particularly important for far IR and THz applications, where no other practical solutions exists beyond particle-type filters. The technology is already sufficiently developed (a photo of a comercial 100mm diameter MPSi LWP filter is shown in Figure 7.8) and the filters were introduced into the market in 2005 by Lake Shore. The most popular applications of MPSi filters todate are FTIR spectroscopy of biological and chemical samples, where the use of LWP filters provides significant improvement in signal-to-noise ratio at long wavelengths, THz research, spectroscopy (where filtering out of the pump laser is required), and astronomy. The filters are currently sold domestically and internationally. This is, to best of the author's knowledge, the first commercially marketed optical product based on porous Si.

7.5 References

[1] Macleod HA, (2001) Thin Film Optical Filters. 3rd Ed., Institute of Physics Publishing.

[2] Armstrong KR, Low FJ, (1974) Far-infrared-filters utilizing small particle scattering and antireflection coatings. Appl. Optics 13:425–430.

[3] Kochergin V, Sanghavi M, Swinehart PR, (2005) Porous silicon filters for low-temperature far IR applications. Proc. SPIE 5883:184–191.

[4] Kochergin V, Föll H, (2006) Novel optical elements made from porous silicon. Review Materials Science and Engineering R, 52:93–140.

[5] Kochergin V, Föll H, (2007) Commercial applications of porous Si: optical filters and components. Phys. Stat. Sol. (c) 4:1933–1940.

[6] Christophersen M, Carstensen J, Feuerhake A, Föll H, (2000) Crystal orientation and electrolyte dependence for macropore nucleation and stable growth on p-type Si. Mater. Sci. Eng. B 69–70:194–198.

[7] Föll H, Christophersen M, Carstensen J, Hasse G, (2002) Formation and application of porous Si. Mat. Sci. Eng. R 39: 93–141.

8

Macroporous Silicon Ultraviolet Filters

8.1 Introduction

Many applications (such as solar blind non-line of sight communications, electrical spark imaging, photolithography, chemical and biological analysis) require optical filters to be used in the ultraviolet ($\lambda < 400$ nm), deep ultraviolet ($\lambda < 300$ nm), or even far ultraviolet ($\lambda < 200$ nm) wavelength ranges. The complexity in fabricating good performance filters for the deep and far UV range grows exponentially for established technologies since materials tend to become absorbant.

No absorption-based filter material is known to have short-pass characteristics with a transparency range down to far UV. While a number of different absorption materials were suggested for band-pass deep UV solar blind filters [1–5], only dye-doped polymer filters [5] from Ofil Ltd. (Israel) have found reasonable commercial success and are currently used in certain demanding applications. It should be noted though that such filters offer generally poor transmission within the pass band (while providing very good levels of rejection) and suffer form quite poor environmental/thermal stability, insufficient for many applications.

Dielectric film technologies for optical coatings employed for ultraviolet applications include deposition of soft, marginally adherent multilayer thin films onto various glasses. The films are soft and lack physical durability; most films are also water-soluble. Usually such films consist of materials such as lead fluoride, cryolite (AlF_6Na_3), zinc sulfide, and so on. Coatings may also contain refractory metal oxides, which are generally more durable. However, standard oxide coatings are optically unstable when exposed to a demanding environment. In addition, soft film filters can be sensitive to temperature and humidity and therefore have relatively limited useable lifetimes. While the deposition technology used for multilayer UV filters is constantly progessing [6], the performance of such filters is often not acceptable for applications where high rejection levels and wide rejection bands are required. For example, stress-related pinholes in multilayer represent one of the major limitations for the practically achievable rejection level.

Another popular class of UV filters, designated as MDM (metal-dielectric-metal) filters [7], are based on a multiple-cavity metal-dielectric Fabry-Perot filter design. MDM filters are comprised essentially of a single substrate of fused silica or quartz, upon which a multilayer coating consisting of two materials, a dielectric (usually cryolite) and a metal (usually aluminum) is deposited. Such filters typically exhibit relatively poor in-band transmission (10–20%) and rejecion levels limited by pinholes (although to lesser extent than all-dielectric filters).

Another conventional approach is filtering by a number of reflective diffraction gratings, which provides fairly good filtering, but at the cost of large, delicate and heavy structures.

Ultraviolet filters based on macroporous Si (MPSi) [8–12] utilize the leaky waveguide transmission mode discussed in Chapter 4 of this book and offer considerable advantages. This filter type will be analyzed here in more detail.

8.2 Theoretical Considerations

As was shown in Chapter 4, at wavelengths below ~ 1 μm an MPSi layer can be considered as an array of leaky waveguides. In the approximation of smooth pure silicon walls, the transparency range of MPSi for the fundamental leaky waveguide mode is limited for longer wavelengths as given by the (approximate) formulas [10] for the –3dB and –20dB attenuation level by:

$$\lambda_{-3dB} = \sqrt{\ln(2) \cdot \frac{d^3}{f_0 \cdot L}} \tag{8.1}$$

$$\lambda_{-20dB} = \sqrt{\ln(100) \cdot \frac{d^3}{f_0 \cdot L}} \tag{8.2}$$

where λ_{-3dB} is a cutoff wavelength at the $-3dB$ level, λ_{20dB} is the rejection wavelength, d is the pore diameter, L is the pore length. For wavelengths longer than 500 nm the factor $f_0 \approx 0.28$.

In other words, a free-standing MPSi layer (or a MPSi membrane) is naturally a short-pass filter. The ratio $\lambda_{-20dB}/\lambda_{-3dB}$, which characterizes the sharpness of the filter edge, becomes a constant $\sqrt{\ln(100)/\ln(2)} \approx 2.6$ that does not depend on the pore size and length in such an approximation. Both the 3 dB cut-off and 20 dB rejection wavelengths scale with $d\sqrt{d/L}$. This means that with deep enough pores, MPSi membranes can exhibit short-pass behaviour at wavelengths far smaller than the pore diameter. For example, at d = 1 μm one can get $\lambda_{-3\text{dB}} = 200$ nm (that is, one tenth of d) if pores with L = 700 μm are used. This pore depth, however, was rather rather challenging from a pore etching point of view but appears to be possible based on new developments as pointed out in Chapter 2. Assuming that a practically achievable aspect ratio is $d/L \approx 100$, the cut-off wavelength becomes $\lambda_{-3\text{dB}} \approx 0.28d$.

As was shown in [9–11], the short-pass filter perfomance of an MPSi layer can be improved by applying thin film coatings to the pore walls to managing the α^{LW} term.

8.2.1 Single-Layer Coatings on MPSi Pore Walls

The simplest realization of a multilayer leaky waveguide array is a free-standing MPSi layer with pore walls uniformly coated by a single layer of transparent dielectric material, as is shown in Figure 8.1. Such a layer will strongly modify the spectral dependences of leaky waveguide loss coefficients by means of constructive and destructive interference of the leaky waveguide mode inside this layer.

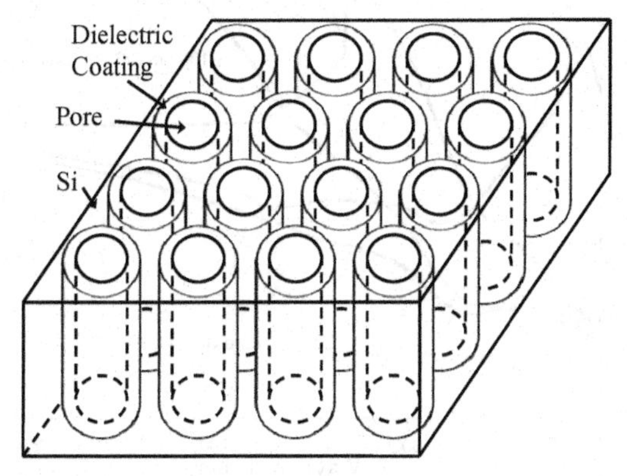

Figure 8.1. Freestanding macroporous silicon uniform pore array with one layer of transparent dielectric material uniformly covering pore walls

To illustrate the effect of the single-layer coating on pore walls, Figure 8.2 shows the calculated spectral dependences of loss coefficients (a) and transmittance through the 50 μm layer (b) of fundamental TE modes for a pure silicon MPSi layer and MPSi layers with pore walls covered by 40, 70 and 100 nm of SiO_2. Pores were assumed to have a near-square cross-section of 1 μm. The modification of the leaky waveguide mode losses in a single-layer coated MPSi array manifests itself as a well-pronounced spectral peak, caused by the suppressed reflection of the leaky waveguide mode from the pore walls at the the wavelength defined by the optical thickness of the pore wall coating.

As illustrated by Figure 8.2, it is possible for a coated MPSi layer to tune the wavelength position of the rejection edge by varying the antireflection layer thickness – just as in the case of common single-layer antireflection coatings. The rejection edge of spectral filters can be tuned in the wavelength from below 200 nm for 40 nm SiO_2 thickness to about 300 nm for 100 nm SiO_2 thickness, while keeping the rejection edge much sharper in comparison to the uncoated MPSi layer and the transmittance within the pass band on about the same level.

While there is a significant similarity in optical effects of single layer coating on a flat surface of silicon and pore walls of MPSi layer, the three-dimensional nature of the macropore causes some well-pronounced differences. For comparison, Figure 8.3 shows calculated spectral dependences of the reflectivity from a pure silicon surface and silicon coated by 40, 70, and 100 nm of SiO_2 at 85 deg. of incidence. The peaks of the loss coefficients of the leaky waveguide modes in Figure 8.2a are quite close to the minima of reflection from the flat silicon surface in Figure 8.3. However, the coincidence is not perfect, and the loss coefficient spectral dependences are considerably more complicated than the reflectivity from the plane interface. The main reasons for such a difference are the spectral variations of the leaky waveguide mode propagation constant, and the non-planar character of the electromagnetic wave in the leaky waveguide mode.

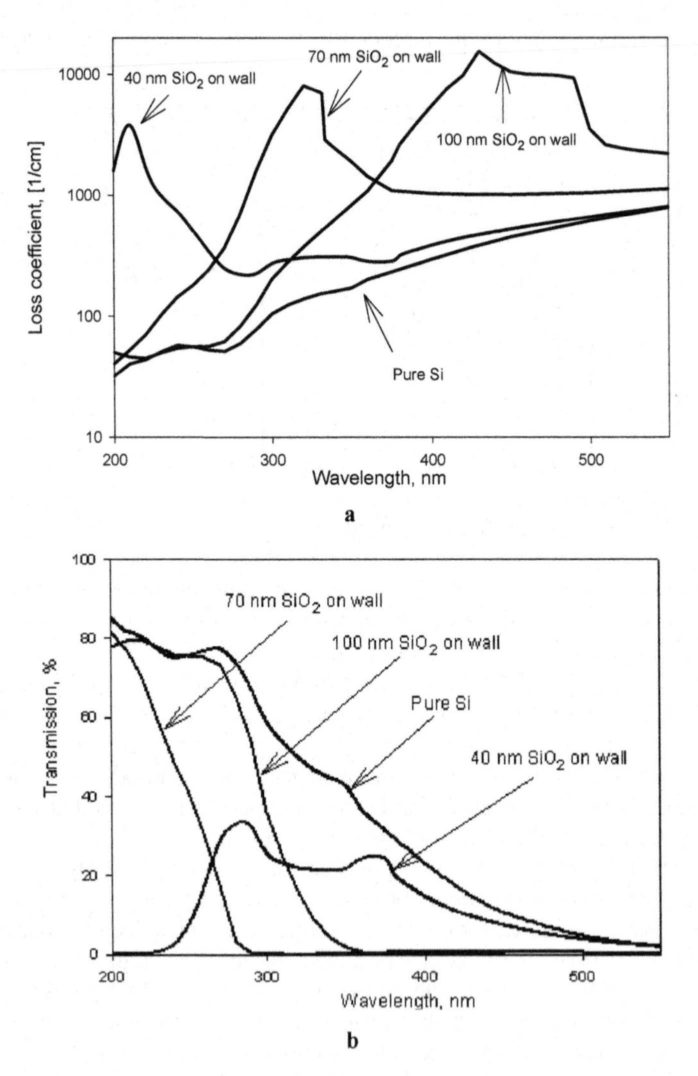

Figure 8.2. Calculated spectral dependences of loss coefficients **a** and of transmittance through the 50 μm layer **b** of the fundamental TE modes for the pure silicon MPSi layer and

the MPSi layer with pore walls covered by 40, 70, and 100 nm of SiO_2. Pores were assumed to have a near-square cross-section of 1 μm (after [9])

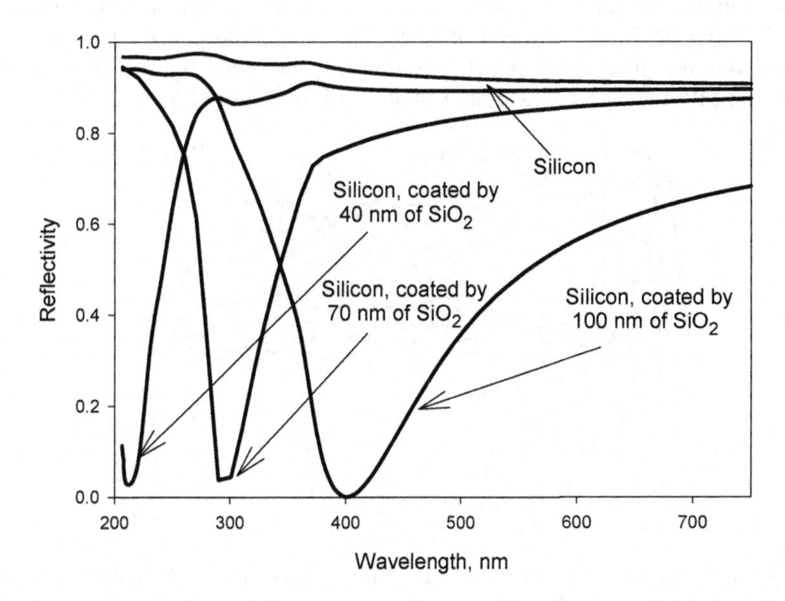

Figure 8.3. Calculated spectral dependences of the reflectivity from a pure silicon surface and silicon coated by 40, 70, and 100 nm of SiO_2 at 85° of incidence (after [9])

The short-pass filters made of single-layer coated MPSi membranes may have some commercial perspectives in not very demanding applications. However, just as with single-layer interference coatings, the design freedom for the filtering characteristics of MPSi layers is very limited. Multilayer coatings of the pore walls can provide significantly higher levels of control of the transmission properties of macroporous silicon layers.

8.2.2 Multilayer Coatings on MPSi Pore Walls

The drawing of an MPSi layer with pore walls covered by several layers of different materials is given in Figure 8.4. As for a single layer coating on the pore walls, the function of multilayers covering the pore walls is to modify the spectral dependence of the leaky waveguide modes loss coefficient ($-\alpha_0^{LW}(\lambda)$). However, unlike a single-layer coating, which can be considered only as an antireflection layer on the top of high refractive index silicon substrate (if this layer is transparent), the multilayer coating can be arranged as a high-dielectric reflector inside the pass band or as a wide-band antireflection coating inside the rejection band of such an MPSi filter.

To illustrate the effect of a multilayer coating on MPSi transmission characteristics, Figure 8.5a gives the calculated spectral dependences of loss coefficients for fundamental TE modes, of an uncoated MPSi layer, and of an MPSi layer having pore walls covered with 70 nm of SiO_2 (similar to the case discussed in relation to Figure 8.3) plus a five-layer coating. Pores were assumed to have a near-square cross-section of 1 μm. The structure of the five-layer coating was as follows: {silicon pore wall / 48 nm of Si_3N_4 / 23 nm of TiO_2 / 59 nm of SiO_2 / 33 nm of Si_3N_4 / 116 nm of SiO_2 / air (inside the pore)}. The multilayer coating was designed as a dielectric mirror for the wavelength of 250 nm (i.e., within the MPSi layer pass band) and an angle of incidence of 85°.

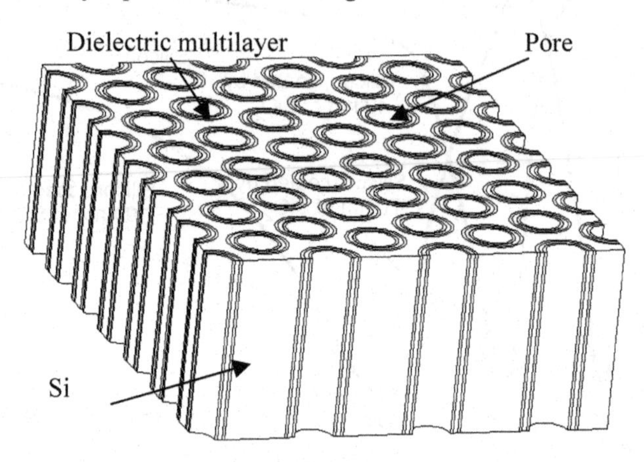

Figure 8.4. Freestanding macroporous silicon uniform pore array with dielectric multilayer uniformly covering pore walls

It is illustrated in Figure 8.5 that the five-layer dielectric coating on the pore walls supresses the leaky waveguide mode losses within the pass band of the MPSi layer more than order of magnitude compared to the single-layer coating discussed in paragraph 8.2.1. At the same time the leaky waveguide mode losses inside the rejection band of an MPSi layer coated with the five-layer high-reflectance coating exceed those of an uncoated MPSi layer by an order of magnitude and are at the same level as that of the single-layer coated MPSi. The short wavelength slope of the loss coefficient peak is much sharper than that of both uncoated and single-layer coated MPSi. These advantageous features allow considerably thicker layers of MPSi to be used, which makes rejection within the rejection band of the MPSi layer considerably higher, while the rejection edge is considerably sharper than that of uncoated and single-layer coated MPSi.

Calculated transmittance spectra through the 200 μm layer for fundamental TE modes of an uncoated MPSi layer and an MPSi layer with the same five-layer coating are shown in Figure 8.5b. One can see that for an MPSi layer with five-layer coated pore walls the transmission edge is sharper by at least 10 times compared to an uncoated MPSi layer. Moreover, the transmission efficiency inside the pass band of such an MPSi layer is about two times higher than the transmission efficiency of an uncoated MPSi layer having the same thickness.

As is illustrated by Figure 8.5b, the width of the pass band for a multi-layer coated MPSi membrane is limited; unlike the case of the uncoated MPSi layer it is expected to be transmissive down to far and extreme UV wavelength ranges. This is caused by the limited width of the high-reflectance band of the multilayer coating. It is possible to increase the width of the high-reflectance band of the coating and thus of the pass-band of the MPSi layer by increasing the number of layers in multilayer, or by increasing the refractive index contrast within the multilayer.

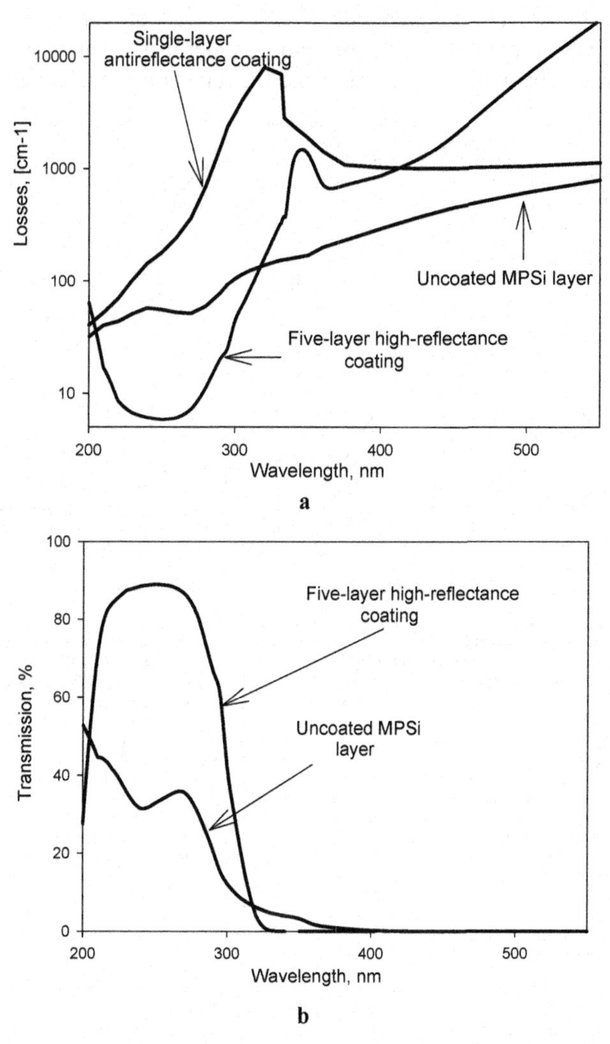

Figure 8.5. Calculated spectral dependences of losses coefficients **a** and transmittance through 200 μm layer **b** of fundamental TE modes for uncoated MPSi layer and MPSi layer with pore walls covered by 70 nm of SiO_2 and five-layer high-reflectance at 250 nm coating (after [9])

One of the most attractive properties of MPSi-based UV filters is the very strong level of rejection that can be obtained over a wide spectral range with very few layer coatings. As an example, Figure 8.6 gives the calculated spectral dependences of the transmission through the MPSi membrane with different numbers of layers in the pore wall coating. The pores were assumed to be 350 μm deep and originally 1.5 μm in diameter (i.e., the pore diameter decreases with the increase of the thickness of the dielectric coating on the pore walls). As follows from Figure 8.6b, even for a single-layer coated MPSi filter the level of rejection can exceed five orders of magnitude over the rejection range. With a higher number of layers on the pore walls (or with a higher aspect ratio of the pores) the rejection edge can be made even sharper and the rejection even deeper.

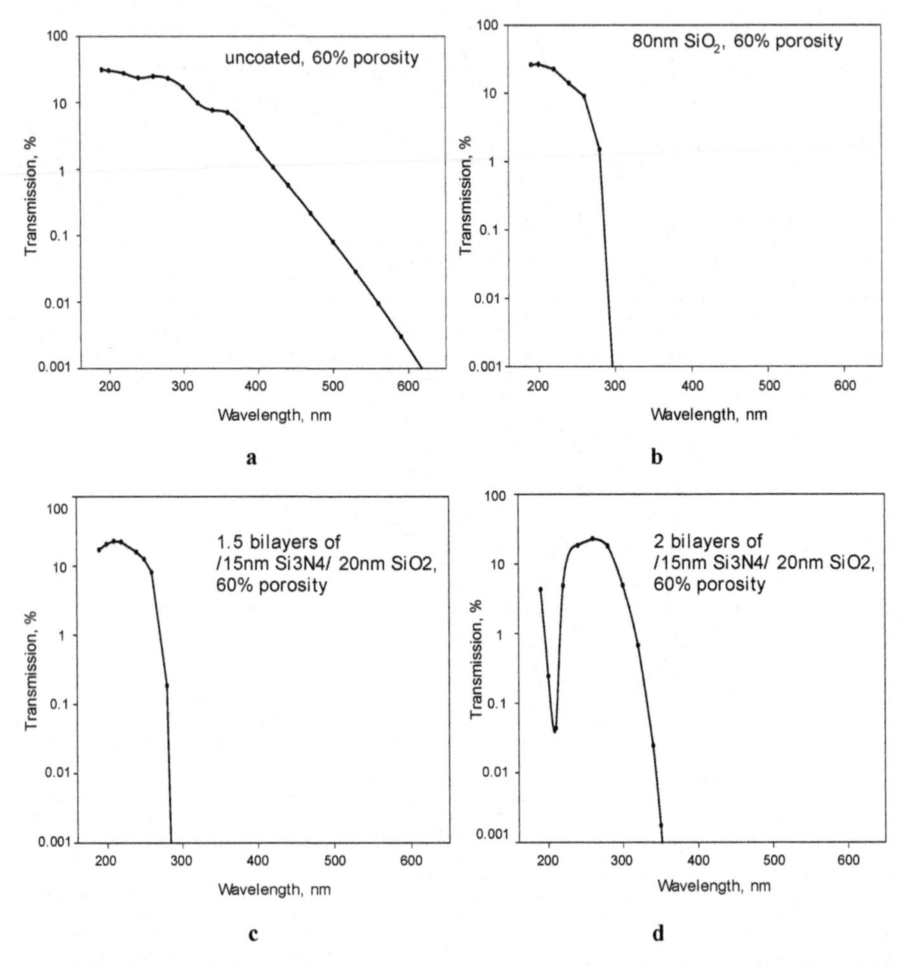

Figure 8.6. Calculated spectral dependences of the transmission through the MPSi membrane with different numbers of layers in the pore wall coatings. The transmission is on a logarithmic scale (after [12])

Another interesting property of MPSi filters is that decent levels of transmission within the pass band can be obtained even if some of the layers in the multilayer show absorption. In the calculations for Figure 8.6 the dispersions of the real and imaginary parts of the refractive indices of Si, SiO_2 and Si_3N_4 were taken into account. One can see that while the Si_3N_4 absorption edge is located approximately at the 290 nm wavelength, substantial transmission in the deep UV is still possible.

Another important feature of MPSi UV filters is the capability of achieving a virtually unlimited rejection band on the longer wavelength side. The MPSi layer itself (uncoated or coated with dielectric multilayers), in addition to the pass band in the UV, also shows pass-bands throughout the near IR and IR (starting from approximately 1100 nm). This transmission band was explained [9] as a consequence of an additional channel of transmission through the MPSi array, which is due to the transparency of Si in the IR. Coating of the interpore areas of MPSi layer surfaces with a metal layer of sufficient thickness can supress the coupling efficiency of light into the Si core waveguides, thus reducing IR transmission channels by several orders of magnitude. By doing that one can obtain a truly short-pass or band-pass deep UV filter: rejection of the wavelengths longer than the rejection edge would be very strong (6 orders of magnitude or more) up to the limit of the waveguide transmission channel (usually in the far IR).

Let's consider now some practially important designs of MPSi UV filters. Frst, let's consider filters for the most demanding application, so-called solar-blind filters, where the very strong rejection (over ten orders of magnitude) throughout all the visible and part of the near IR is needed. The following pore structure was found to theoretically provide sufficiently deep rejection, a sufficiently sharp edge and acceptable levels of transmission. Listed radially inward from the pore wall we have:{Si pore wall / 57 nm HfO_2 / 73 nm SiO_2 / 7 nm HfO_2 / 111 nm SiO_2 / 18 nm HfO_2 / 33 nm SiO_2 / 52 nm HfO_2 / 91 nm SiO_2 / 800 nm air-filled pore}. The original pore diameter is 1680 nm.

Some calculated transmission spectra through MPSi filter with such a basic coating structure with 300 μm thick porous layers are given in Figure 8.7. Figure 8.7a shows the calculated transmission spectra on a linear scale, while Figure 8.7 gives the calculated transmission spectra on a logarithmic scale. The inserts in both figures provide magnified portions of the plots in a UV region of spectrum. The curves show variants of the basic structure as follows:

- The A curves on both figures correspond to the calculated transmission through the MPSi layer with perfectly uniform pores and pore wall coatings along the pore depth, and 150 nm metal coating on both sides of the porous layer. No tapering of the pore walls was assumed (hence there is room for improving the transmission through the filter).
- The B curves correspond to the same structure without metal coating of the surfaces and with 5% random nonuniformity of the layers on the pore walls (quite a reasonable number as preliminary experiments with ALD pore wall coating have shown).
- The C curve corresponds to the MPSi layer with metal coating and with 5% nonuniformity assumed.

- The D curve corresponds to the same parameters as the C curve but now for a MPSi layer with 400 μm thickness, 5% nonuniformity and metal coating on the surface.

According to the calculations, such a design not only offers a filter performance considerably exceeding the most demanding specifications, but also permits reasonable latitude in MPSi layer parameters. The relatively small number of layers (eight) (layers for the pore wall coating not only can provide for reasonably low costs for these filters, but also promises to avoid some problems, typical for multi-layer interference filters such as growth of a columnar structure and scattering of light at a large number of always imperfect interfaces.

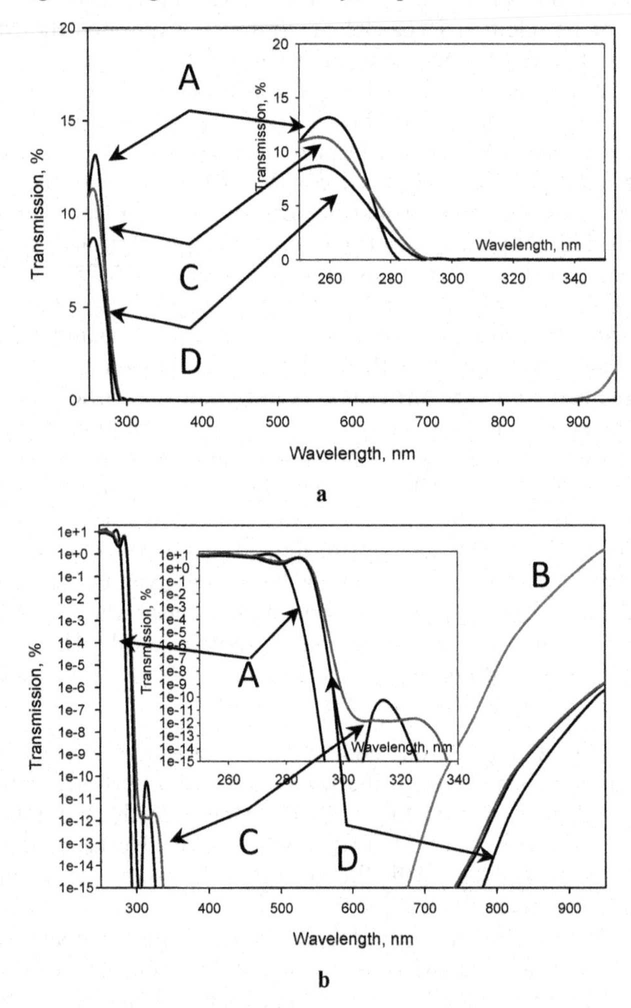

Figure 8.7. Numerically calculated transmission spectra through an MPSi membrane with an eight-layer pore wall coating. The transmission (y-axis) is shown in **a** on a linear scale and in **b** in logarithmic units. The inserts show the magnified portions of the spectra in UV region. See text for differences in curves

Besides the short-pass type of filter, it is possible to fabricate a band-pass filter based on coated MPSi layers as well. The peculiarity of the MPSi filter structure are such that, in order to exhibit narrow bandpass behavior, the coating on the pore walls should act as a narrow band reflector at an almost glancing angle of incidence (depending on the pore geometry at angles between 80° and 86°). As it is well known, in order to form such a reflector one needs to use a substantial number of layers of materials with adequately close indices of refraction. This is because the width of the band in this case is determined by the index contrast – the higher the contrast, the wider the band. This calls for the "engineering" of the refractive index of the layers in the multilayer stack, since the choice of transparent, environmentally stable and depositable materials is quite limited in the deep UV range. As will be shown later there is considerable potential for multilayers of "engineered" materials with some of the deposition techniques mentioned.

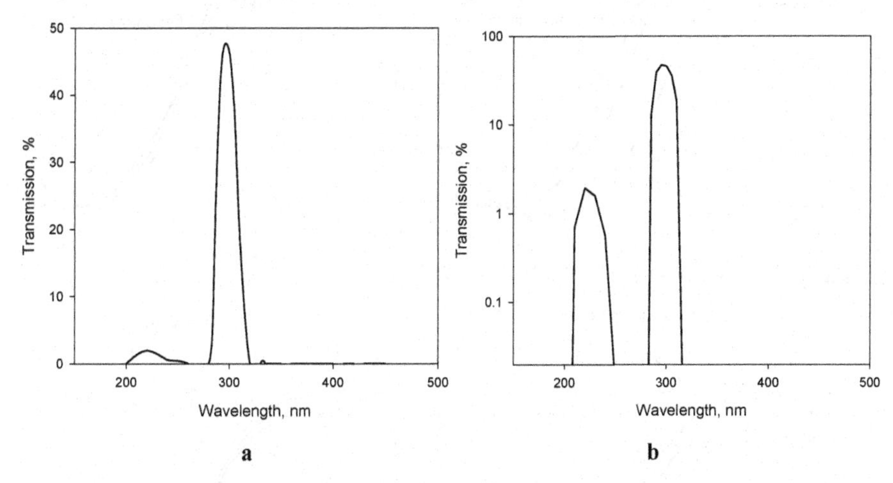

Figure 8.8. Calculated transmission spectra through an MPSi membrane with a 19-layer pore wall coating designed as a narrow bandpass filter with a band centered at 300 nm. **a** Transmission on a linear scale. **b** Transmission on a logarithmic scale with more detail (after [13])

As an example, Figure 8.8 gives the calculated transmission spectra of an MPSi membrane with a 9.5-bilayer pore wall coating. It is designed as a narrow bandpass filter with the band centered at 300 nm. In the calculations the refractive indices of the layers were assumed to be 1.75 and 1.9. The dispersion of the dielectric constants in the pore wall coating was neglected. Such refractive indices are close to those of an Al_2O_3 and Al_2O_3/TiO_2 nano-laminated stack, discussed later in this chapter. In the calculations an MPSi membrane with an aspect ratio of 150 was used (such an aspect ratio is well within the limits of the technology). The narrow bandpass transmission of such a structure is clearly demonstrated.

One should note that in addition to a quite high level of transmission within a reasonably narrow band, such a structure should exhibit high levels of rejection at wavelengths above the center of the band despite a parasitic transmission peak at ~220 nm resulting from a theory neglecting absorption. However, this peak should be strongly suppressed in reality since it is located at wavelengths below the TiO_2 absorption edge (i.e., the TiO_2 portion of a multilayer will be opaque).

MPSi layers with coated pore walls will provide the following advantages over ordinary short-pass filters: the steepness of the rejection edge will be increased (due to multiple reflections that the light-waves experience during propagation through leaky wave-guide), the rejection level will be higher by many orders of magnitude, and the transmission spectrum will be independent of the angle of incidence due to the independence of the coupling/outcoupling and filtering processes. Although the mechanism of improving the transmission spectra of the MPSi layer with multilayer-coated pore walls is interference-based, the MPSi layer will not suffer from the disadvantages of the usual interference filters, such as the dependence of the spectrum on the angle of incidence [9].

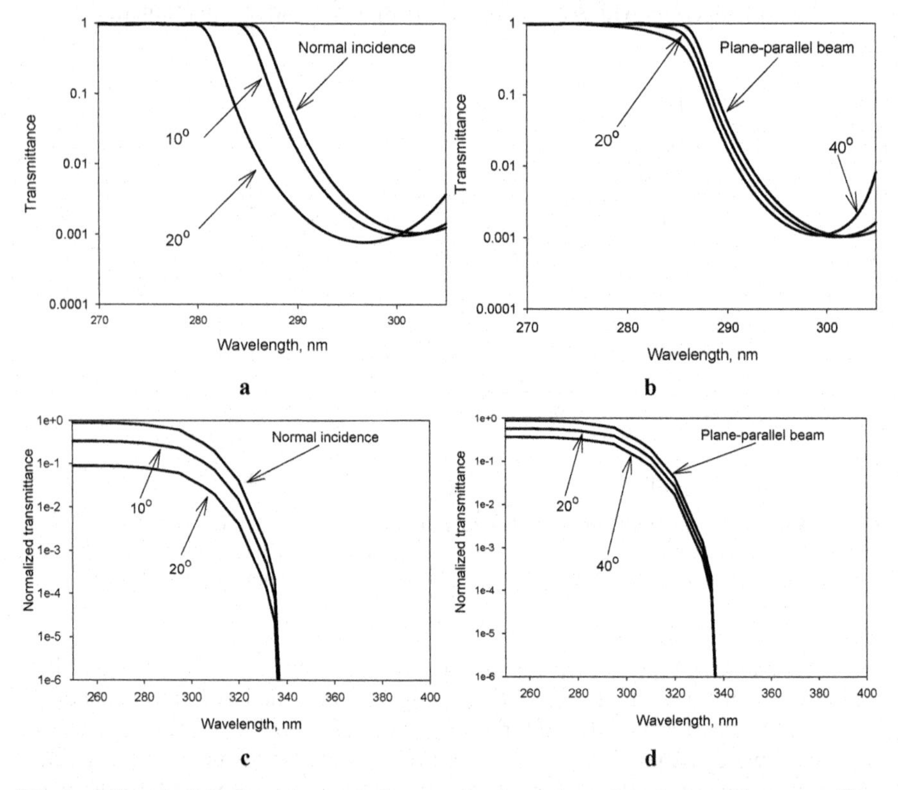

Figure 8.9. a–b Calculated transmission spectra through an interference short-pass filter **a** and through a five-layer coated MPSi layer **b** for different angles of incidence. **c–d** Calculated transmission spectra through an interference short-pass filter **c** and through a five-layer coated MPSi layer **d** for different diversions of Gaussian beam (after [9])

Figure 8.9a shows the calculated transmittance spectra through the typical interference short-pass filter for normal incident, 10° and 20° tilted plane-parallel beams. It illustrates the wavelength shift of the pass-band edge position common to all interference edge filters. Figure 8.9b presents the normalized transmittance (maximum transmittance will be defined by the particular filter structure and can vary from ~ 20% to 75%) spectra through an MPSi layer with a five-layer coating for normal incident, 20° and 30° tilted plane-parallel beams. By comparing these plots we can conclude that a coated MPSi layers can be used as short-pass filters at different angles of incidence (± 20° at least). A similar effect can be expected for convergent/divergent beams: Figure 8.9c illustrates the calculated transmittance spectra through the interference short-pass filter for normal incident beams with different convergences (plane-parallel beam (0-convergence angle) and Gaussian beams with 20° and 40° convergence angle). It illustrates the degradation of both the band-edge shape and out-of-band rejection, common to the interference short-pass filter. For a comparison, Figure 8.9d presents the normalized transmittance spectra through an MPSi layer with 5-layer coating for 0, 20° and 40° convergent normally incident Gaussian beams.

8.2.3 Wide Field-of-View Design of MPSi UV Filters

The MPSi UV filter design presented in the previous paragraphs offers absence of the spectral shifts or degradation of the shape of the passed or blocked spectral bands with the angle of incidence, thus permitting operation in tilted and divergent light beams as long as the angle of incidence does not exceed the acceptance angle of the pore structure. If the pores are left unfilled as in the previously discussed designs (see Figures 8.1 and 8.4), such a filter can be transmissive down to the far UV and extreme UV spectral range. However, as will be shown in the experimental section of this chapter, the acceptance angle of such filters is quite narrow (between 10 and 20 degrees, depending on MPSi layer parameters).

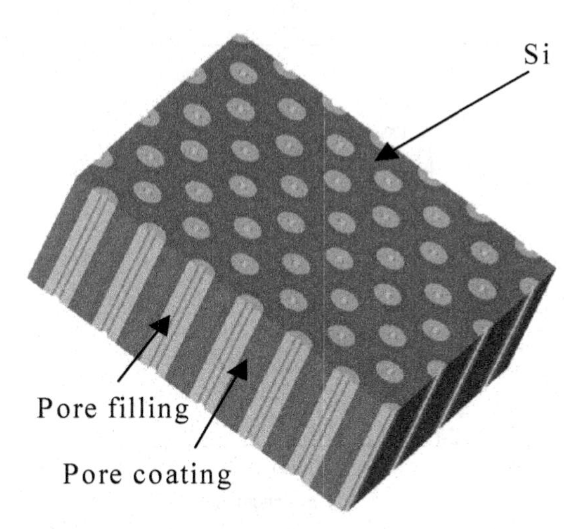

Figure 8.10. Drawing of the pore structure for the omnidirectional wide field-of-view UV filter design, also defining the terms "pore coating" and "pore filling"

Omnidirectional UV filters with wider field-of-view are also possible in the design utilizing pores almost completely filled with transparent material (as shown in Figure 8.10). Calculations show that such structures are possible and the field-of-view can be indeed enlarged by a factor, which is approximately equal to the refractive index of the pore-filling material. For deep UV filters the pore filling material choice is limited to the materials that are sufficiently transparent at these wavelengths. For example, according to preliminary calculations, HfO_2 (refractive index 2.3) or Al_2O_3 (refractive index 1.7) can provide sufficient transmission while increasing the angular range of the filter by a factor of 2.3 and 1.7, respectively. Such materials can be deposited by, for example, Atomic Layer Deposition (ALD) or Low Pressure Chemical Vapor Deposition (LPCVD).

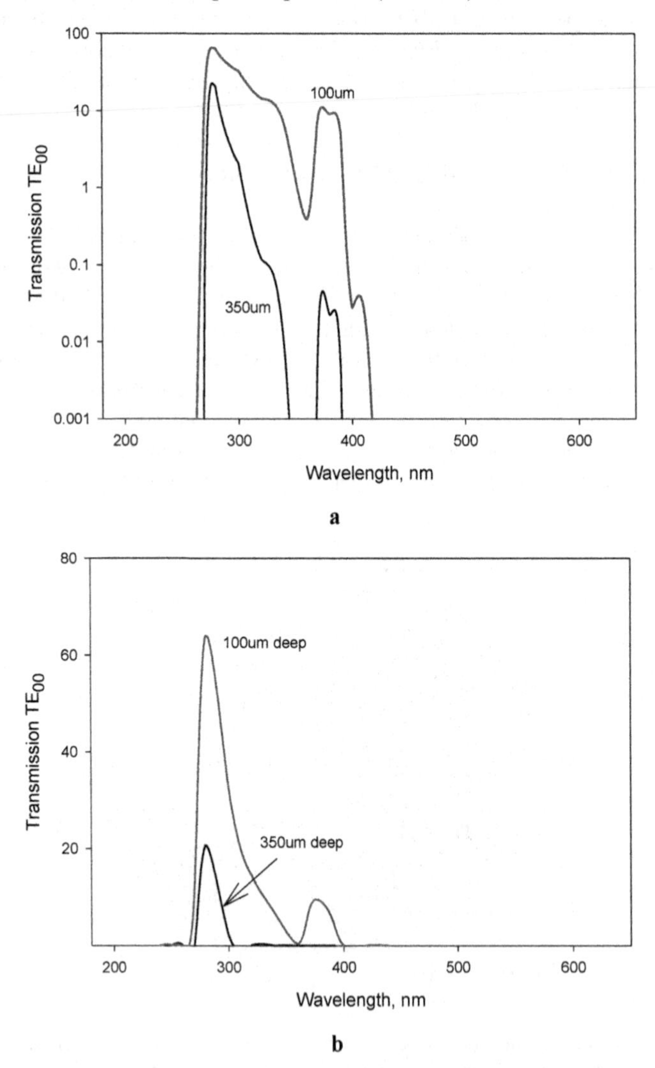

Figure 8.11. Calculated spectral dependences of the transmission efficiency through a TE00 leaky waveguide mode through 100 μm and 350 μm deep MPSi membranes with a 5-layer coating (50 nm SiO_2/14 nm Si) leaving 900 nm diameter after coating. The filling is Si_3N_4 (with 25 nm diameter unfilled area in the center). **a** shows the transmission in logarithmic scale, while **b** represents the linear scale

There are other factors that have to be accounted for while designing the "filled" MPSi UV filters. One of them is that the average refractive index of the coating should exceed the effective refractive index of the leaky waveguide mode, i.e. has to exceed the refractive index of the pore-filling material (otherwise the pore will embody a regular waveguide and no longer a leaky waveguide and no modification of the transmission spectrum will be possible). This limits the possible range of materials to be used for the pore coating and the pore filling. For the case of LPCVD, where the possible materials are limited to polysilicon, silicon dioxide and silicon nitride, it causes the necessity to use polysilicon in the multilayer coating (which is opaque at such wavelength). However, calculations (Figure 8.11) show that even with such materials it is possible to design deep UV filters.

8.3 Fabrication and Results

Formation of MPSi was reviewed in some detail earlier in Chapter 2, so there is no need to repeat it here. The focus of this section will be on the other processes required to make a functional UV filter from an MPSi layer or membrane. First, as mentioned previously, a freshly etched MPSi layer should be made into a free-standing membrane in order to gain any kind of transmittance through it at UV wavelengths. Membrane formation requires removing the non-etched part of the silicon wafer from the back side. This can be done, e.g., by inductively-coupled reactive ion etching (RIE) or by chemical etching in hot KOH solutions [8]. Figure 8.12 shows an SEM image of a cleaved, free-standing MPSi membrane fabricated by back-side chemical etching. In order to protect the pore walls from chemical species during chemical back-etching, a thin protective layer of Si_3N_4 was deposited in this case by a CVD process onto the pore walls, which after the process was removed in hot phosphoric acid.

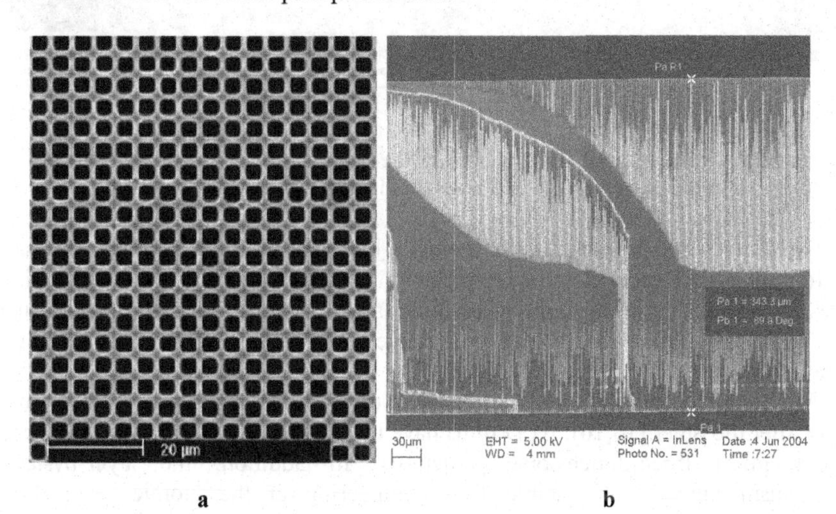

Figure 8.12. SEM images of the MPSi membrane. **a** Top view, **b** Cross-sectional view (after [14])

The next process required to fabricate an UV MPSi filter is pore wall coating. The simplest process for obtaining a single-layer pore wall coating is thermal oxidation of the MPSi membrane. Early results on optical properties of thermally oxidized MPSi layers were reviewed in detail in [9]. In order to obtain multilayers of different dielectric materials on the pore walls, a different technique has to be used. Low Pressure Chemical Vapor Deposition (LPCVD) process was suggested [9] and successfully demonstrated [14] for coating MPSi pore walls. MPSi filters with up to seven layers coating the pore walls were successfully fabricated by a Low Pressure CVD technique by MEMS Precision Instruments, Inc [14]. SEM images of the LPCVD-coated MPSi layer cross-section are given in Figure 8.13, which gives the image of the single-layer silicon nitride-coated MPSi membrane in a), while b) shows the image of the five-layer coated $\{SiO_2/Si_3N_4/SiO_2/Si_3N_4/SiO_2\}$ MPSi membrane. In both cases, a thin layer of polysilicon was deposited by an LPCVD technique after finishing the dielectric coating in order to obtain a better contrast of the SEM images, which is not present in samples used for optical characterization. Samples were polished down from the top for these pictures and the cross-sections shown were made at a depth of ~200 μm down from the original MPSi layer surface.

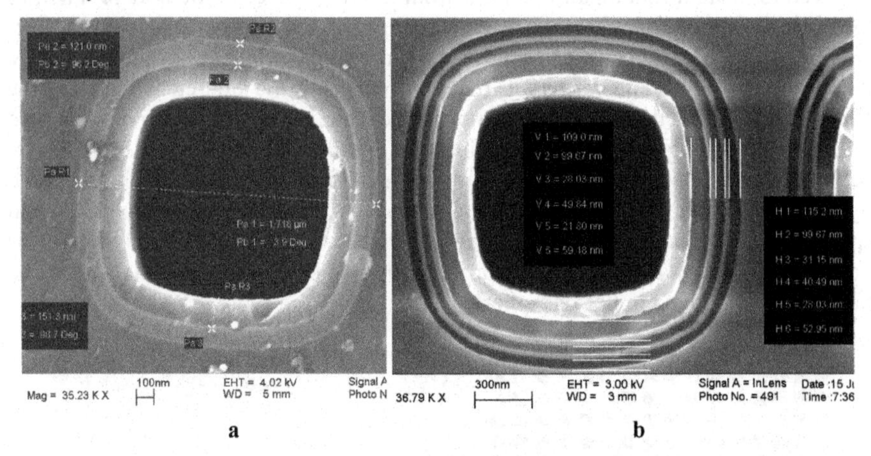

Figure 8.13. Cross-sectional SEM images of the MPSi layer with pore walls coated by: **a** single layer of dielectric material. **b** 5 layers of dielectric materials (after [12,14])

By using LPCVD techniques, it might be difficult to obtain the necessary uniformity of the individual layer thickness' along the pore length. In addition, stresses in the films were found to be quite large, causing various problems (such as a severe sample bowing). An Atomic Layer Deposition (ALD) technique might be a better process for pore wall coating.

ALD is a deposition technique that lays down films sequentially, one atomic layer after the other [15,16]. ALD films have the proven ability to conformally coat the toughest three-dimensional structures. In addition, the layer-by-layer mechanism allows reproducible film engineering at the atomic level. Film engineering is a key process in enabling continuous growth of ALD films over a variety of substrates wherein typically the incompatibility of the film with the substrate is overcome by providing an ultrathin (several monolayers only) interface layer that is compatible with both the substrate and the film. Film engineering has also been proven to modify and improve film properties such as the suppression of the growth of disadvantageous polycrystalline films [17]. Amorphous films are important for MPSi UV filter fabrication, since with amorphous films scattering-related trasmission losses are significantly smaller than with polycrystalline films.

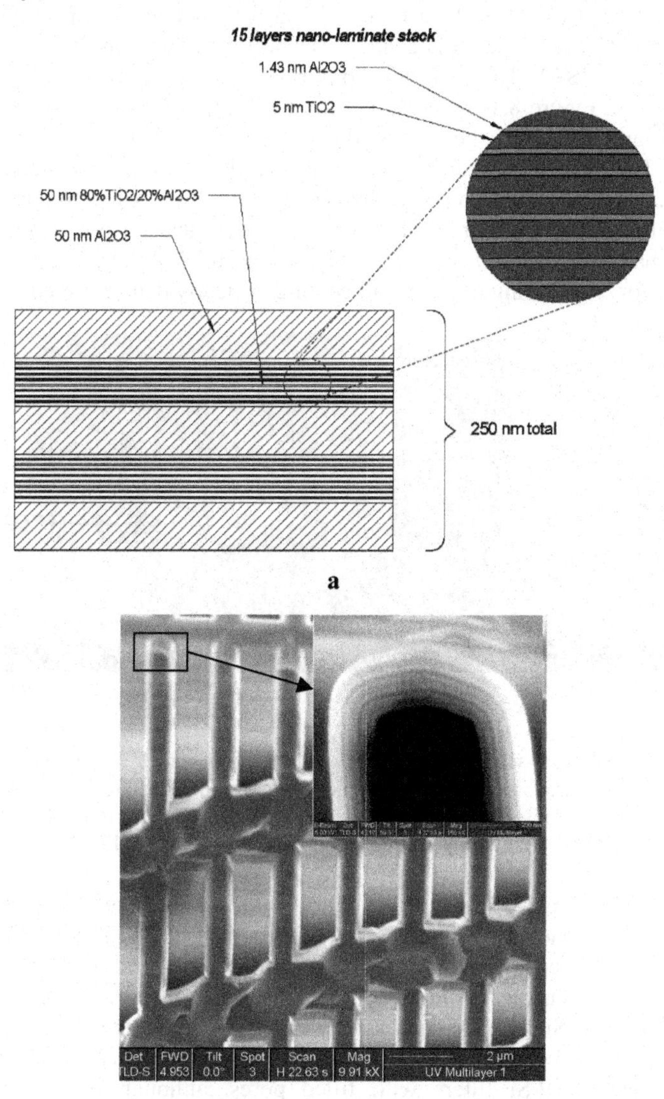

Figure 8.14. **a** Schematic cross-sectional drawing of the pore wall coating structure used in the experiment. **b** Cross-sectional SEM images of an ALD-coated MPSi membrane with an Al_2O_3/TiO_2 multilayer with the structure as in **a** (after [13])

The feasibility of using the ALD technique for MPSi pore wall coatings was successfully demonstrated [12,13]: a 250 nm thick ALD film stack was deposited on and in a MPSi sample using five alternating bilayers consisting of a pure Al_2O_3 layer and an alumina-titanate layer, each 50 nm thick. The individual alumina-titanate layer comprises fifteen nanolaminated layers of 5 nm thick TiO_2 and 1.43 nm thick Al_2O_3 resulting in a 4:1 TiO_2:Al_2O_3 composition ratio that is sufficient to suppress the natural, disadvantageous agglomeration of TiO_2, while achieving a relatively high optical index (see Figure 8.14a for a schematic illustration of the fabricated layer-stack). The ALD coating was performed by Sundew Technologies, Inc. (Colorado, USA). The resulting conformal coating is apparent in the SEM images given in Figure 8.14b, indicating the feasibility of robust and reproducible multilayer growth with the required conformality over challenging MPSi substrates. The SEM picture clearly indicates that the 250 nm thick stack is highly conformal. The high resolution SEM highlights the 5 layer stack, although sample charging effects blur the borders between the layers and do not allow to reveal individual nanolayers of 5.0 nm TiO_2 and 1.43 nm thick Al_2O_3. The feasibility of using ALD for engineering of pore wall coating is clearly demonstrated.

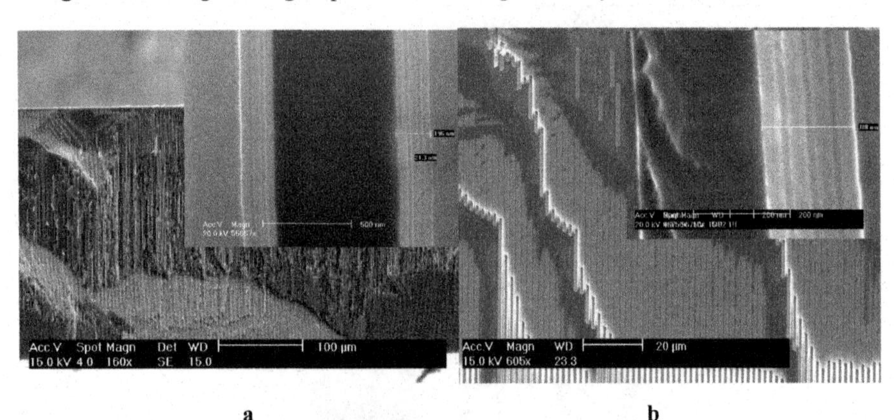

Figure 8.15. SEM cross-sectional images of: **a** Five-layer, and **b** seven-layer TiO_2/Al_2O_3 coating on the pore walls of MPSi membrane. Magnified images are shown as inserts on both figures. Scales on insert in **a** are 500 nm, 196 nm and 21.3 nm, from largest to smallest, respectively. Thickness of multilayer on insert in **b** is 308 nm. After [12,13]

Images of less challenging multilayer pore wall coatings are provided in Figure 8.15. Figure 8.15a shows a 5-layer TiO_2/Al_2O_3 coating, while Figure 8.15b shows a 7-layer TiO_2/Al_2O_3 coating. High uniformity and good structural properties of the coating are clearly visible. It should be noted that other pore-wall coating materials (such as HfO_2 and SiO_2) were successfully deposited by ALD onto the pore walls as well. It is believed that ALD will be a technique of choice to fabricate wide acceptance angle MPSi filters with filled pores, although this is still to be experimentally demonstrated.

8.4 Optical Testing

Most of the theoretical predictions on optical properties of MSPi UV filters were experimentally validated [9–13]. As a confirmation of the theoretically predicted high rejection level over the wide spectral band, the transmission spectrum through an MPSi membrane with an adequately high aspect ratio (200) is given in Figure 8.16a. One can see that rejection in uncoated samples exceeds 5 "O.D. units" (O.D. = optical density) from 300 nm and at least up to 800 nm (i.e. over part of the UV range, over all the visible part of the spectrum, and over at least part of near IR range). The transmission in the near IR was in the order of the noise level of the IR detector (which was much noisier than the UV-sensitive detector). Metal coating of one or both surfaces of the MPSi membrane (a 49 nm thick ion-beam sputtered Ta layer) enables even more spectacular rejection levels. As follows from Figure 8.16a, the level of IR transmission through the MPSi membrane (up to 3300 nm) was suppressed down to the spectrometer detection limit, while the UV transmission band stayed unchanged (all curves exactly match).

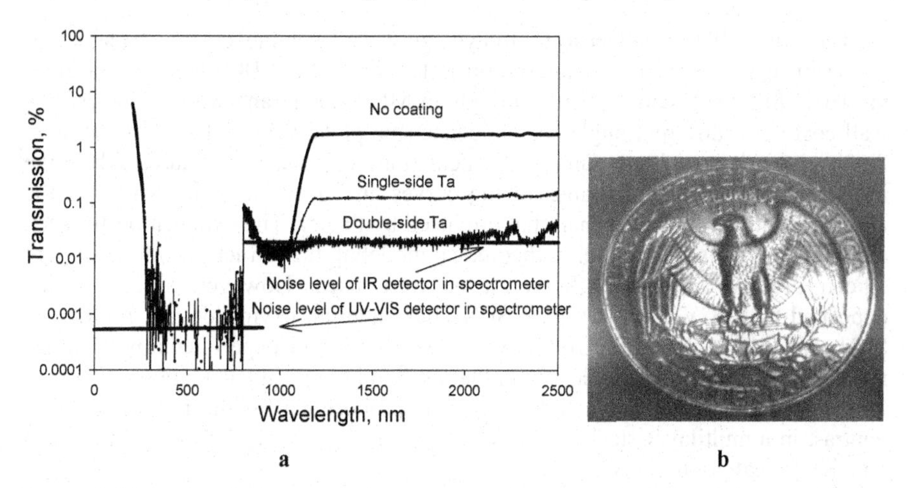

Figure 8.16. **a** Far field transmission spectrum through an MPSi membrane with and without metal coatings on one or both surfaces. **b** Photo of a quarter taken through the MPSi filter (so the CCD array is effectively in the far field). After [14]

Imaging properties of the MPSi UV filters are demonstrated in Figure 8.16b. This figure presents the image of a coin taken through the filter. Figure 8.17a demonstrates that the (nano)roughness of the pore walls is an important factor for the achievable transmission levels.

The high achievable levels of for the near- and far-field transmission efficiency are illustrated by experimental plots given in Figures 8.17a and 8.17b. One can see that in the deep UV (300 nm and below) over 35% far field transmission and up to 90% near field transmission are possible. It should be noted that while the level of near field transmission is close to the practically achievable value, there is still a room for improvement on the far-field transmission of MPSi filters.

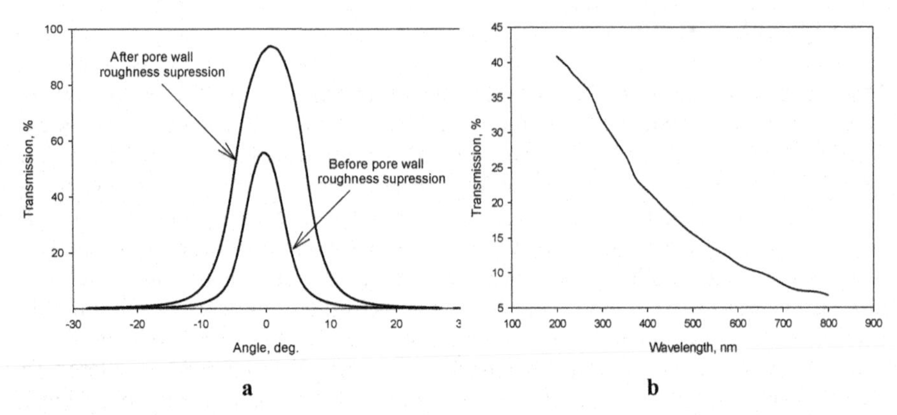

Figure 8.17. a Experimental angular dependences of the near-field transmission through an MPSi UV filter at 248 nm wavelength. **b** Far field transmission spectrum through another MPSi membrane. After [9,12]

The effect of the dielectric multilayer pore wall coating on the transmission spectrum of the MPSi membrane is illustrated in Figure 8.18. Figure 8.18a gives the far field transmission spectra through an MPSi membrane with a 5-layer pore wall coating at different angles of incidence (the plots were scaled), while Figure 8.18b gives the result of numerical calculations of such a structure. While the measured and calculated shapes of the curves are fairly close, the width of the experimental peak is lower than that predicted by theory. The explanation for such a discrepancy is fairly simple: in theoretical modeling the refractive indices of pure silicon dioxide and pure silicon nitride were used. However, silicon nitride deposited by LPCVD is not a pure Si_3N_4, but rather an oxynitride (i.e. mixture of Si_3N_4 and SiO_2). Thus, the refractive index contrast in experimentally obtained multilayer was substantially less than that used in numerical calculations. It is well known that the width of the Bragg peak is proportional to the refractive index contrast in a multilayer stack, which explains the discrepancy between experiment and theoretical predictions.

Figure 8.18 not only confirms the major theoretical prediction that transmission through an MPSi membrane can be increased over a wavelength band due to the Bragg reflection phenomenon in pore wall coatings, but also confirms the omnidirectionality of the filters. Such a property is very important for narrow bandpass applications.

8.5 Conclusions

Macroporous silicon UV filters represent a new type of optical filter and promise a number of important advantages over the conventional interference or absorption-type filters, such as omnidirectionality, very high and wide rejection range, truly short-pass (down to far UV) transmission. While the theoretical understanding of principles of operations of MPSi UV filters is well developed [9,10,12] and major predicted optical properties are experimentally validated [9,11,12,14], more research and development is still required to make marketable products based on this technology. With active development of both macroporous silicon etching [18] and continuous advancements of the coating technologies [16] though, the development of such filters can be expected in the near future. Moreover, porous layers of different semiconductors (such as InP [19]) may provide a basis for similar type optical filters as well.

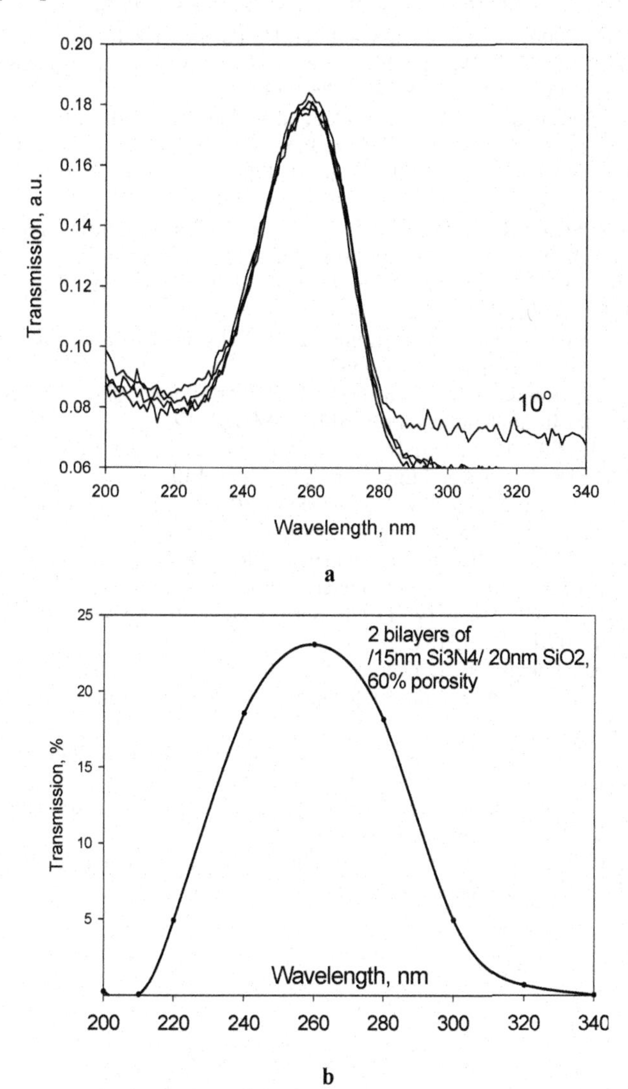

Figure 8.18. a Far field transmission spectra through an MPSi membrane with a 5-layer SiO_2/Si_3N_4 coating on the pore walls, taken at different angles of incidence and scaled in order to visualize the omnidirectionality of the filter. The 4 curves correspond to angles between 0° and 10°. **b** Numerically calculated transmission spectrum through an MPSi membrane with 2 SiO_2/Si_3N_4 bi-layers covering the pore walls. After [13,14]

8.6 References

[1] Kraushaar RJ, Ward KB, (1982) Thallium doped potassium iodide filter. US Patent # 4,317,751.
[2] Kraushaar RJ, Ward KB, (1986) Solar blind ultraviolet filter. US Patent # 4,597,629.
[3] Singh NB, Partlow WD, Strauch S, Stewart A, Jackovitz JF, Coffey DW, Mazelski R, (1998), Crystals for ultraviolet light filters. US Patent # 5,837,054.
[4] Haaland PD, (2000) Solar blind optical filter. US Patent 6,126,869.
[5] Lindner MB, Elstein S, Wallace J, Lindner P, (1998) Solar blind bandpass filters for UV imaging devices. Proc. SPIE 3302:176–183.
[6] Potter JM, (2005) Advanced Ultra-Violet and Visible Narrowband Interference Filter Technology. Proc. ESTC-05 Conference, June 2005, Baltimore, Maryland, USA.
[7] Macleod HA, (2001) Thin Film Optical Filters. 3rd Ed., Institute of Physics Publishing.
[8] Lehmann V, Stengl R, Reisinger H, Detemple R, Theiss W, (2001) Optical shortpass filters based on macroporous silicon. Appl. Phys. Lett. 78:589–591.
[9] Kochergin V, (2003) Omnidirectional Optical Filters. Kluwer Academic Publishers, Boston, ISBN 1-4020-7386-0.
[10] Avrutsky I, Kochergin V, (2003) Filtering by leaky guided modes in macroporous silicon. Appl. Phys. Lett. 82: 3590–3592.
[11] Kochergin V, Sanghavi M, Swinehart PR, (2005) Porous silicon filters for low-temperature far IR applications. Proc. SPIE 5883:184–19.
[12] Kochergin V, Föll H, (2006) Novel optical elements made from porous silicon. Review Materials Science and Engineering R, 52:93–140.
[13] Kochergin V, Sneh O, Sanghavi M, Swinehart PR, (2005) Macroporous silicon deep UV filters. Proc. ESTC-05 Conference, Baltimore, Maryland, USA.
[14] Kochergin V, Christophersen M, Swinehart PR, (2004) Macroporous Silicon UV filters for space and terrestrial environments. Proc. SPIE 5554:223–234.
[15] George SM, Ott AW, Klaus JW, (1996) Surface chemistry for atomic layer growth. J. Phys. Chem. 100:13121–13131.
[16] Sneh O, Clark-Phelps RB, Londergan AR, Winkler JL, Seidel TE, (2002) Thin film atomic layer deposition equipment for semiconductor processing. Thin Solid Films 402:248–261.
[17] Ritala M, Leskela M, (1997) in: H.S. Nalwa, (Ed.), Handbook of Thin Film Materials, Vol. 1, Chapter 2, p. 103, Academic Press, San Diego
[18] Föll H, Christophersen M, Carstensen J, Hasse G, (2002) Formation and application of porous Si. Mat. Sci. Eng. R 39: 93–141.
[19] Föll H, Langa S, Carstensen J, Christophersen M, Tiginyanu IM, (2003) Review: Pores in III-V Semiconductors. Adv. Materials, 15:183–198.

9

Polarization Components for the UV Range

9.1 Introduction

Optical polarization components are used in numerous applications that include photography, liquid crystal displays, polarimetry, astronomy, defense, sensing and photolithography, to name just a few. There are many types of polarizers and polarization components currently available. Polarizers based on metal wires, birefringence in crystals and on specially designed dielectric multilayers are among the widest spread (although the variety of known polarizers is by no means limited to these technologies). While polarizers oriented for the visible and near IR wavelengths ranges perform well, that is not the situation in the UV, especially for deep and far UV. For example, inexpensive plastic polarizers are not useable at UV wavelengths since the plastics are not sufficiently transparent. The high-end UV polarizers (such as those based on birefringence (Glan-Thompson) or on multilayers (cube beam-splitters)) are more expensive than the same types designed for the visible and near IR range. In addition, their performance degrades substantially at the shorter wavelengths of light. For example, deep UV polarizers designed to work at 248 nm offer an extinction of just 100–1000, compared to 10^5 for visible polarizers, but demand a high price even for a small working area. In addition, the performance of almost all current types of polarizers capable of performing at deep UV and shorter wavelengths are strongly dependent on the angle of incidence of light on the polarizer. For example, tilting the incident light beam away from normalincidence, the extinction degrades considerably and/or the preferred polarization may rotate. Far UV polarizers are available only by special order, are extremely expensive, and offer fairly poor performance compared to polarizers for the visible spectrum.

Departing from conventional technology, an entirely new concept for a UV optical polarizing material was proposed recently [1] that promises solutions to the problems described above. These new polarizers consist of ordered, macroporous silicon (MPSi) with the pore walls covered by dielectric multilayers, similar to the MPSi-based UV filters considered in detail in Chapter 8. The polarizing behavior in this case originates from the elongated shape of the pores, which provides for the selection of one preferential linear polarization from the transmitted light as described in this chapter.

9.2 Theoretical Considerations

The MPSi considered so far in this book has either a random structure (as with LWPF) or for defined structures still a radial-symmetrical pore shape (either square or round). Because of this symmetry, the structures did not alter the polarization state of the transmitted beam and were functional only as optical filters. Albeit classical macropores have a circular or quadratic shape, it is possible to generate rectangular slit shapes by arraying the pores in lines combined with a chemical overetch [2]. The so-called Ottow technique [3] allows even to structure porous arrays in arbitrary shapes, but necessitates a complex post-etch structuring process. Recently it was shown [4], that MPSi arrays can be fabricated in a much richer variety of shapes even in a one-step-etch-process, including MPSi arrays with long asymmetrical pores.

A freestanding MPSi membrane with rectangular or elliptic pores exhibits polarizing behavior in transmission, as was shown by Kochergin et al. [1]. This behavior originates from the higher propagation losses of the leaky wave-guide modes for electric field vectors directed along to the shorter pore axes (see Figure 9.1) compared to those of leaky waveguide modes with their electric field vectors directed along the longer pore axis. This statement can be supported by an estimation of losses in a waveguide formed by metallic-like boundaries surrounding a dielectric core (e.g. vacuum or other transparent relatively low refractive index material).

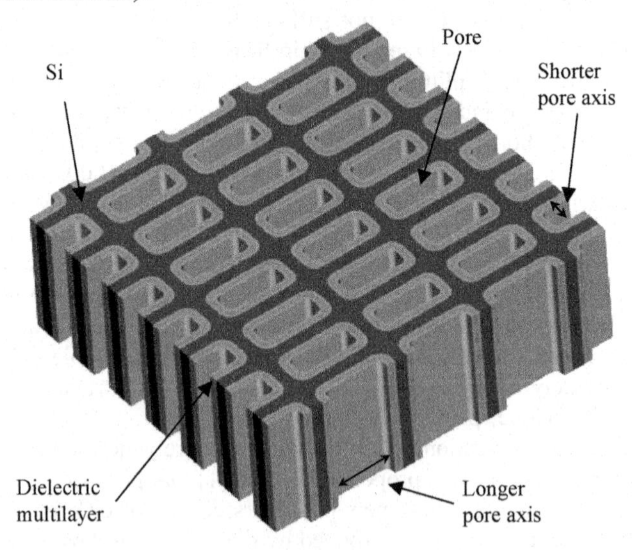

Figure 9.1. Schematic drawing of an MPSi-based UV polarizer (after [1])

The theoretical treatment of such a material is quite similar to that in Chapter 4. Denoting the length of the longer pore axis with a and the length of the shorter pore axis with b and keeping all other abbreviations, it follows that at short wavelengths ($\lambda < 294$ nm) the difference in the loss coefficients of the modes polarized along the shorter and longer pore axes can be estimated from Equation 4.2 as

$$\Delta\alpha \approx (N+1)^2 \lambda^2 \frac{n}{n^2+\kappa^2}\left(\frac{1}{b^3}-\frac{1}{a^3}\right) \tag{9.1}$$

Similarly, in the wavelength range $\lambda > 294$ nm, from Eq (4.3) it follows that

$$\Delta\alpha \approx (N+1)^2 \lambda^2 \frac{1}{\sqrt{n^2-1}}\left(\frac{1}{b^3}-\frac{1}{a^3}\right) \tag{9.2}$$

Both equations (9.1) and (9.2) predict a strong increase in the polarizing behavior of the MPSi layer with the increase of the difference in the length of the pore axis, number of the mode, and the wavelength. However, it should be noted that Equations 9.1 and 9.2 are limited by the loss coefficient of the TM mode, which is generally considerably larger than that of the TE mode. This manifests itself in a saturation of the anisotropy at some specific value of the difference in pore axis a–b. The important property of an MPSi membrane with elongated pores is the exponential increase of the extinction (defined as usual as the ratio of the transmittances for the two orthogonal polarizations) with the depth of the pores. Hence, high extinction is possible for deep pores over a wide wavelength range, including far UV.

An important factor that affects the applicability of the polarizing material to real applications is not only the extinction but also the transmission of the preferred polarization. Hence, not only $\Delta\alpha$ is of importance, but also $\Delta\alpha/\alpha$. At realistic values of the pore parameters for uncoated MPSi layers, this ratio is limited to about an order of magnitude, meaning that it is impossible to achieve high levels of extinction (in excess of 1000) at reasonable levels of transmission of the preferred polarization. This problem can be solved by applying once more a dielectric multilayer coating to the pore walls, as in the UV filters discussed in Chapter 8. A schematic drawing of such an MPSi-based polarizer is given in Figure 9.1. It consists of a freestanding MPSi membrane with elliptical or rectangular pores coated with a dielectric multilayer having a specially designed structure. Figure 9.2 gives a numerically calculated plot of the transmission spectral dependences through a 200 micrometer thick MPSi layer, as described below. In this example, the pores were assumed to have 1.3×0.7 micrometer cross-sections and a 5-layer coating on the walls. The coating had the following structure: {Silicon pore wall / 10 nm SiO_2 / 33 nm Si_3N_4 / 60 nm SiO_2 / 41 nm Si_3N_4 / 95 nm SiO_2}. As suggested by theory, a wavelength band of over 80 nm is predicted with better than 5000:1 extinction between orthogonal polarizations of the transmitted light while a high level of transmission for the preferred polarization (in excess of 50%) is still preserved. It should be noted that the center wavelength of the band could be tuned over a wide range (through the far UV to near IR spectral ranges) by scaling the dimensions of the pores and tuning the structure of the multilayer coating on the pore walls. The extinction of the polarized material can also be increased by either increasing the aspect ratio of the pores or by increasing the number of layers on the pore walls.

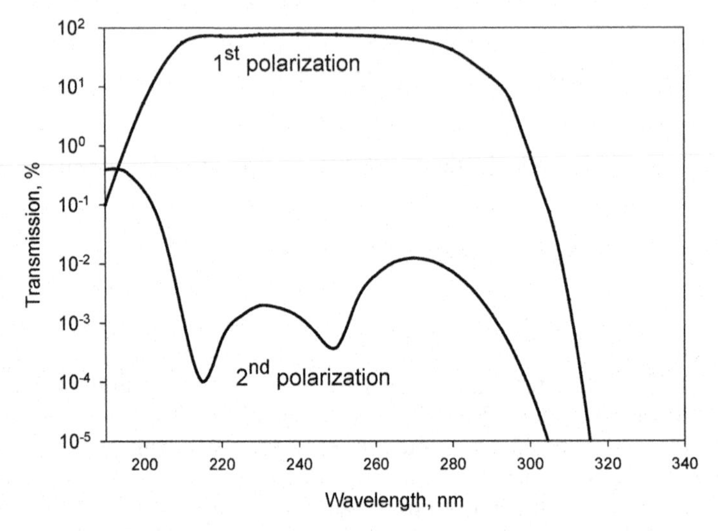

Figure 9.2. Calculated spectral dependences of the transmission of light with two orthogonal polarizations through an MPSi structure similar to that in a). 1st polarization corresponds to the electric field vector aligned along the longer pore axis, while the 2nd is along the shorter pore axis (after [1])

9.3 Experimental Results

For optical demonstration of the polarization behavior, an MPSi membrane with elongated pores was fabricated. The fabrication of the random MPSi array is essentially similar to that reviewed in detail in previously except that the etch pits that were intentionally made with an elongated shape. The SEM image of the surfaces of MPSi layer used for optical testing is given in Figure 9.3. It should be noted that the design goal (see Figure 9.1a) was not met yet and more optimization of the electrochemical etching process is required. The pore structure of the sample that was fabricated looks like that shown schematically in Figure 9.3b. Meanwhile, in the framework of the German BMBF research project "Phokiss" established to develop a novel MIR gas sensor, deep elongated groove etching has been further improved, in the context to develop a superior coupling facette to a photonic crystal. While normally, grooves decompose after etching of some ten microns into individual pores [4], an optimized process makes it possible to etch well-defined grooves up to depths of 170 microns and more.

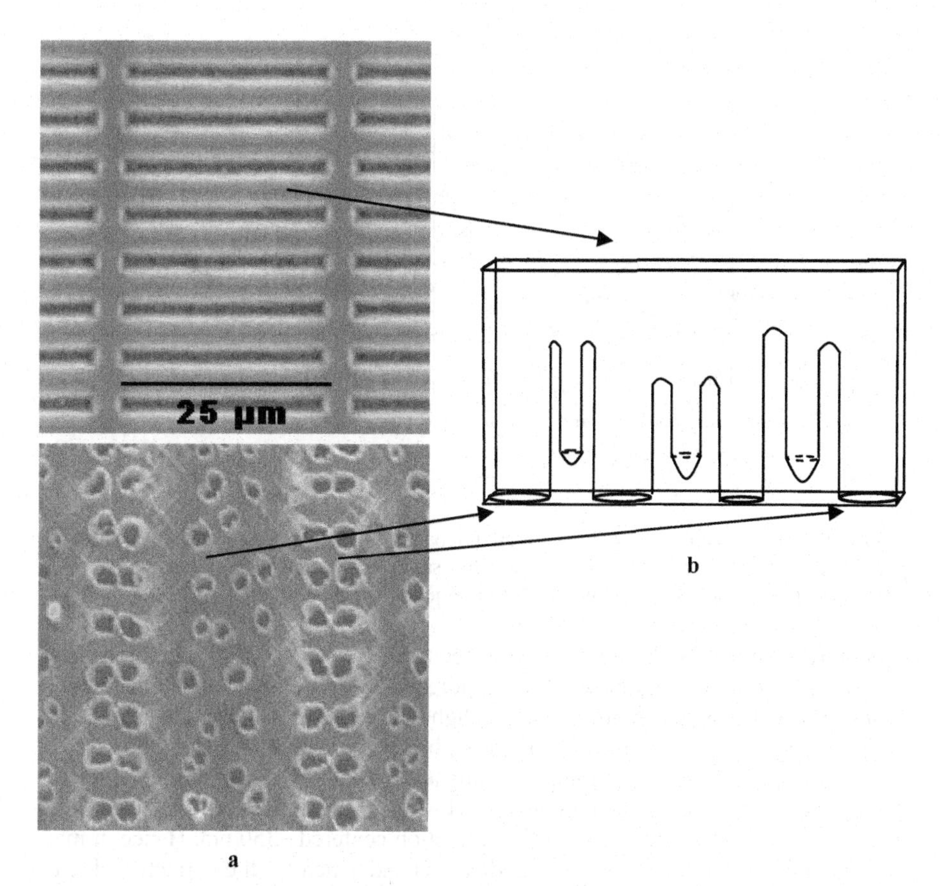

Figure 9.3. SEM images of the MPSi arrays. **a** Top and bottom view (after "opening" the MPSi membrane by RIE) of original surface. **b** Schematic drawing illustrating the structure of the MPSi layer shown in **a**. After [1]

To prove the theoretical predictions, MPSi membranes (oxidized and uncoated) were tested in a VARIAN CARY 500 UV-VIS-near IR spectrophotometer. A commercial polarizer optimized for the visible wavelengths was used to analyze the polarization. Figure 9.4a shows transmission spectra of the uncoated MPSi membrane for two states of polarization of the incident light. It should be noted that the nonuniformity of the transmission for each polarization ("peaks" at ~300 nm and ~400 nm) are caused by the spectral non-uniformity of the extinction of the commercial polarizer. Spectral dependence of the extinction of this sample is given in Figure 9.4b. One can see that the sample indeed showed some polarization behavior, as predicted by theory. Another conclusion that can be made from Figure 9.4b is that the extinction of the MPSi membrane increases with the wavelength approximately parabolically, in agreement with theoretical predictions.

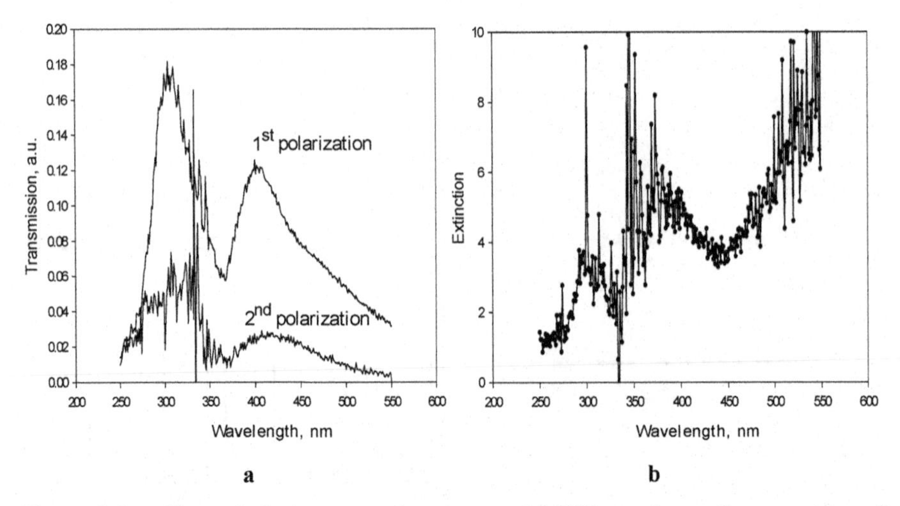

Figure 9.4. a Transmission spectra of an uncoated MPSi membrane for two states of polarization of incident light (notations are the same as introduced regarding Figure 9.2. **b** Spectral dependence of the extinction of the same sample as in **a**. After [1]

Figure 9.5a shows the transmission spectra of the same MPSi membrane, but with a thin layer of SiO_2 grown on the pore walls by thermal oxidation for two states of polarization of the incident light. The spectral dependence of the extinction in this case is given in Figure 9.5b. The transmission peaks and valleys due to the oxide presence on the pore walls are masked by much stronger spectral variations of the commercial polarizer. However, the oxide on the pore walls manifests itself as a strong peak of the extinction centered ~350 nm. Hence, it may be concluded that the theoretically predicted enhancement of the extinction due to dielectric coating of the pore walls was demonstrated.

The experimentally achieved values of extinction for the MPSi-based polarization component were far less spectacular than those predicted by theory. This is well understandable, however, since the quality of the tested MPSi layer was quite far from the design goals and in reality only the upper part of the sample was responsible for obtaining the polarization properties.

It should be also noted that the deep and far UV polarizing properties of the material were not tested due to the absence of a reference polarizer with both good transmission and good extinction within these spectral ranges. The commercial polarizer used showed a transmission cut-off at about the 260 nm wavelength. However, since the experimental results confirming the main predictions of the theory, polarizing properties of better quality MPSi material in the deep and far UV ranges can be expected.

9.4 Conclusions

Macroporous silicon UV polarization components promise significant improvement in performance in the UV and especially deep UV range over the existing technologies. Theoretical understanding of the principle of operation of such components is well developed to date, however, there are still significant experimental efforts required to move MPSi polarizers to the market. While experiments made to date qualitatively confirm the theoretical predictions, the quantitatively measured high levels of extinction are yet to be achieved. However, with recent improvements in electrochemical etching (see Figure 9.6) the authors are quite confident that high-performance MPSi-based UV polarizers can be fabricated in the near future. Moreover, polarization plates with complex spatial polarization profile can be realized with MPSi technology as well due to the unique tenability of MPSi morphology, as illustrated by an SEM image in Figure 9.6.

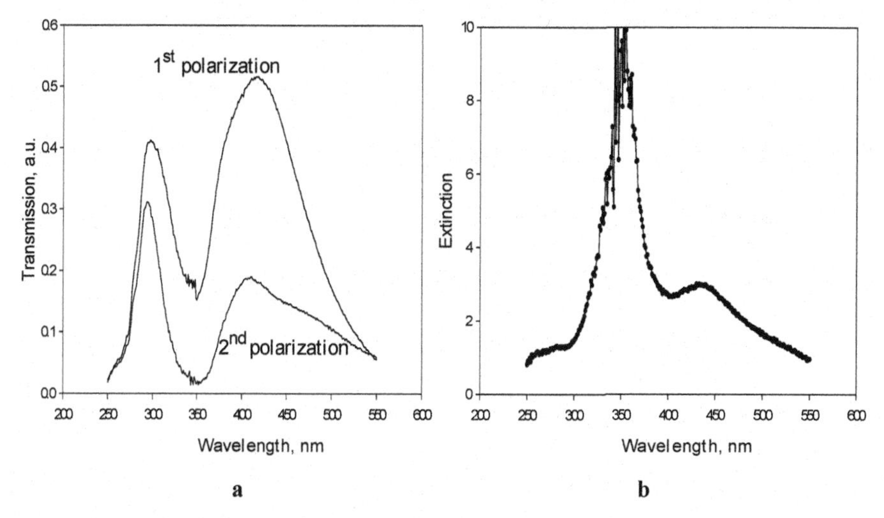

Figure 9.5. **a** Transmission spectra of the MPSi membrane with a thin layer of SiO_2 grown on the pore walls for two states of polarization of incident light. **b** Spectral dependence of the extinction of the same sample as in **a**. (after [1])

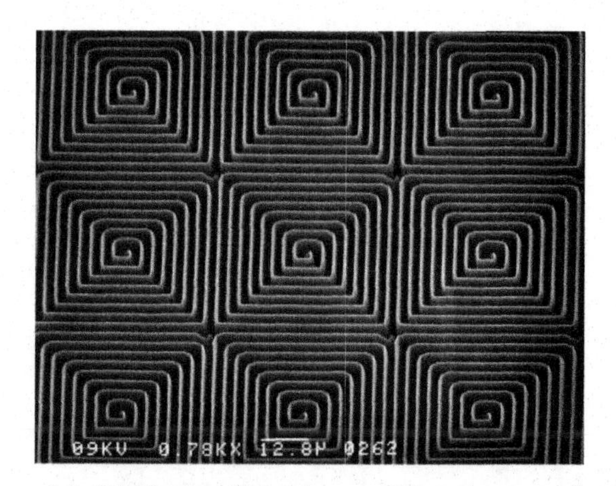

Figure 9.6. SEM image of Macroporous Silicon structure (after [4])

9.5 References

[1] Kochergin V, Christophersen M, Swinehart PR, (2004) Macroporous Silicon-based polarization components. Proc. SPIE, 5515: 132–141.

[2] Lehmann V, Rönnebeck S, (2001) MEMS Techniques Applied to the Fabrication of Anti-Scatter Grids for X-Ray Imaging. in Technical Digest of the 14th IEEE International Conference on Micro Electro Mechanical Systems, 84, Interlaken (2001).

[3] Ottow S, Lehmann V, Föll H, (1996) Processing of three-dimensional microstructures using macroporous n-type silicon. J. Electrochem. Soc. 143:153–159.

[4] Barillaro G, Nannini A, Piotto M, (2002) Electrochemical etching in HF solution for silicon micromachining. Sensors and Actuators A 102:195–201.

10

Retroreflection Suppression Plates

10.1 Introduction

Infrared sensors have been called "the eyes of the digital battlefield" [1]. Military applications dominate the requirements today, especially for IRFPAs (infrared focal plane arrays, which are optical sensors placed at the focal plane of an IR optical system such as a camera, night vision system or night gun sight). In addition to the many military applications for IR systems such as target acquisition, search and track and missile seeker guidance, there is also great potential for IR systems in the commercial marketplace. IR systems can enhance automobile and aircraft safety, medical diagnosis and manufacturing quality and control. Uncooled long wavelength infrared range (LWIR) FPA's, i.e. those not requiring cryogenic cooling equipment, have enjoyed significant recent advances, making them lighter, simpler and easier to install and maintain. These sensors are now being considered for many "Future Combat System" platforms to meet target acquisition, navigation and surveillance requirements. However, the reflectivity of current uncooled LWIR sensor technology is unacceptably high due to high refractive index semiconductors used in the FPA: If they are used in an otherwise "invisible" airplane, they can be seen, e.g. by CW search lasers as illustrated in Figure 10.1. In military applications it is therefore very important to minimize the signature of sensors in certain wavelength regions without compromising performance. Thus, it is highly desirable to reduce the reflectivity of uncooled imaging sensors.

This goal can be accomplished by several approaches. For example, the angle of incidence of the incoming radiation may be diverted optically so as to minimize the retroreflection, or the incoming image may be modified by use of optical or image processing techniques to render it out of focus, followed by re-image for display. Both of these approaches may provide sufficient results, but the former requires additional optics and a longer light path, thus making the FPA optical system bigger, heavier and more complex, while the latter has the disadvantage of leading to a decrease of the FPA's resolution and lengthens the time for signal processing. A third method to accomplish the suppression of retroreflection from an FPA is to place a special retroreflection suppression plate in the vicinity of the FPA surface, which will not disturb the image transmission in the direction towards the FPA, but will strongly and uniformly scatter all the retroreflected light. Realization of such a retroreflection suppression plate on the basis of porous semiconductor technology was suggested and demonstrated in [2] and will be reviewed in this chapter.

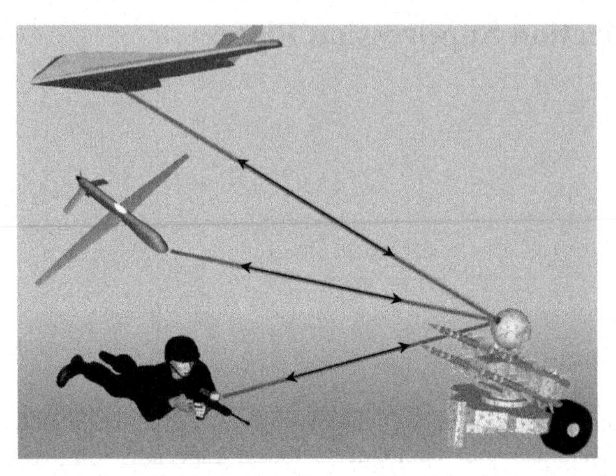

Figure 10.1. Schematic drawing illustrating the possible detection of friendly forces by an enemy vehicle employing laser scanning because of from uncooled LWIR FPAs (after [2])

10.2 Principle of Operation

Supression of retroreflection with porous silicon component is based on the severe difference between the far- and near-field transmission efficiencies in random MPSi arrays in the leaky waveguide transmission mode. If, for example, a LWIR FPA is placed into the near field of a random MPSi array as shown in Figure 10.2, the light incident on the upper surface of the MPSi plate will be mostly transmitted through massively parallel leaky waveguides to the FPA array, while light retroreflected from the FPA array will be scattered during reemission from the non-coherent (disordered) leaky waveguides. Therefore the retroreflection will be effectively suppressed. The main difference between this retroreflection suppression plate design and prior art light diffusing elements is that the coupling of light into such a plate introduces very little back-scattering (less than 1% in the tapered macropore design), while the majority of scattering occuring during the outcoupling process. This is true no matter from which side of the plate light is incident because the reflection from the plate itself is minimal and no scattered light is transmitted through the plate. For example, if a prior-art diffuser (such as described in [3] and references therein) were to be used instead of an MPSi-based plate, the scattering at the upper surface of the diffuser would erase or strongly blur the image (assuming the thickness of the diffuser exceeded the FPA pixel-to-pixel distance, which is almost always the case). Another important advantage of the porous semiconductor retroreflection suppression plate is the wide range of angles and wavelengths over which it suppresses retroreflection [2].

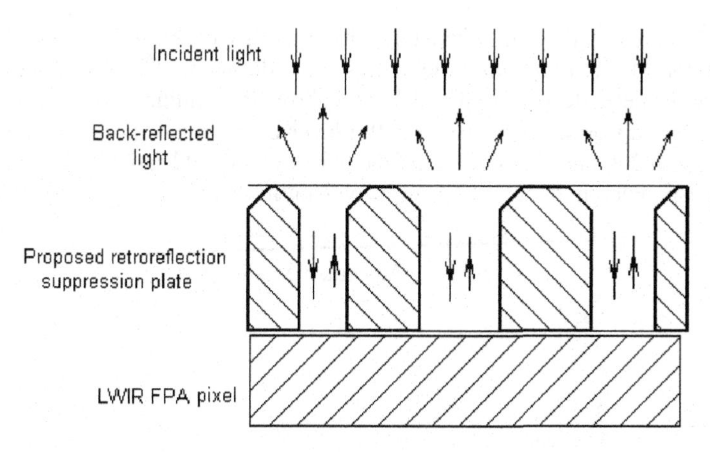

Figure 10.2. Reduction of the retroreflection from uncooled LWIR FPA with the help of a MPSi membrane. A random free-standing MPSi layer is placed on the top of LWIR FPA in near-field (after [2])

To maximize the performance of MPSi-based retroreflection suppression plates, the following conditions should be satisfied:

1. The near-field transmission efficiency should be maximized.
2. The retroreflection suppression should be maximized (i.e., the far-field transmission should be minimized for a wide range of angles and wavelengths).
3. The image captured by the FPA should be not distorted.

In order to design the retroreflection suppression plate one has to define first the spectral band of interest,i.e. the wavelengths at which the retroreflection needs to be suppressed while the near field transmission have to be maintained at high levels. Since the leaky waveguide mode of transmission through the MPSi layer has to be utilized, the results of the theory presented in Chapters 4 and 8 will be used. Let's, for example, estimate the pore dimensions and other MPSi layer properties that would provide an optimized performance in the 8 μm to 12.5 μm long wavelength infrared range (LWIR, according to US military specifications). As was shown in Chapters 4 and 8, both the 3 *dB* cut-off and 20 *dB* rejection wavelengths of uncoated MPSi arrays used in leaky waveguide modes scale proportionally to $d\sqrt{d/L}$ (where d is the diameter of the pore and L is the length of the pore). This means that for the LWIR range and an MPSi thickness of 500–700 μm (common thickness of silicon wafers as supplied by vendors), the pore diameter d should be around 30 μm. However, fabrication of such large pores is a significant challenge at least for now. Fortunately, the transparency range of the MPSi layer may be enlarged and the transmission efficiency within the transparency range may be improved while keeping the pore sizes at more reasonable levels if a thin metal coating is introduced into the pore walls. This is illustrated by Figure 10.3 where the numerically calculated near-field transmission (corrected for porosity) through a 500 μm-thick MPSi layer is shown for uncoated and coated pores with 20 μm diameter. The feasibility of ALD-coating of the pore walls of an MPSi array with dielectrics was shown in Chapter 8, while the feasibility of metal deposition by ALD was shown in a number of references (see e.g., [4]). It is apparent from that an MPSi layer consisting of pores having 20 μm diameters and 200 nm of Au or Ag on the pore walls would satisfy the requirement of the transparency in the 7.5–12.5 μm wavelength range.

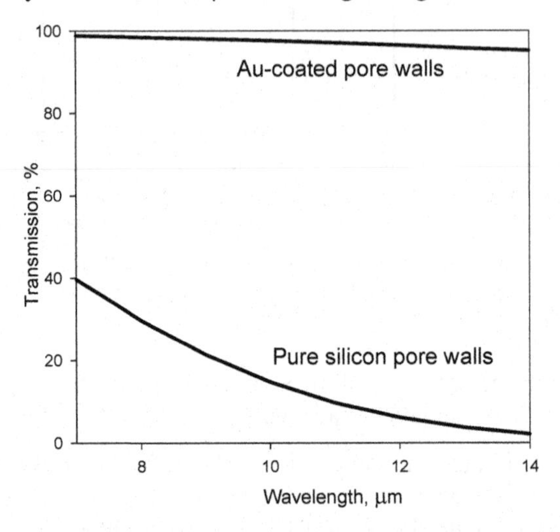

Figure 10.3. Calculated near-field transmission (corrected for porosity) through a 500 μm-thick MPSi layer with pore diameters of 20 μm. Cases for pure silicon MPSi and for a 200 nm-thick Au layer uniformly covering the pore walls are shown (after [2])

Maximization of retroreflection supression can be translated into the following conditions for porous silicon layer:

1. Low far-field transmission across a wide spectral range (from UV to CO_2 laser wavelengths, since such wavelengths may be employed by the enemy's detection equipment), and
2. Low far-field transmission across reasonably wide angles.

The condition 1) is automatically satisfied if condition 2) is satisfied (provided the pore size is smaller than the wavelength of light by at least five times. This is the case here since the proposed pores are 20 μm in diameter while the CO_2 wavelength is just 10 μm). The condition 2) basically means that not only long-range order in the pore positions should be avoided, but that short-range order should also be minimized. The latter requirement is necessary to avoid rings of diffracted light, which might be observed by an enemy. Porous silicon layers with random pores optimized by proper adjustment of the electrochemical etching parameters (such as temperature and current during the pore nucleation and reorganization stage) [5] can meet these conditions. Alternatively specially constructed random arrays of pores, as, e.g. used in [6] may be employed. The first option is preferred since it reduces the number of necessary processing steps.

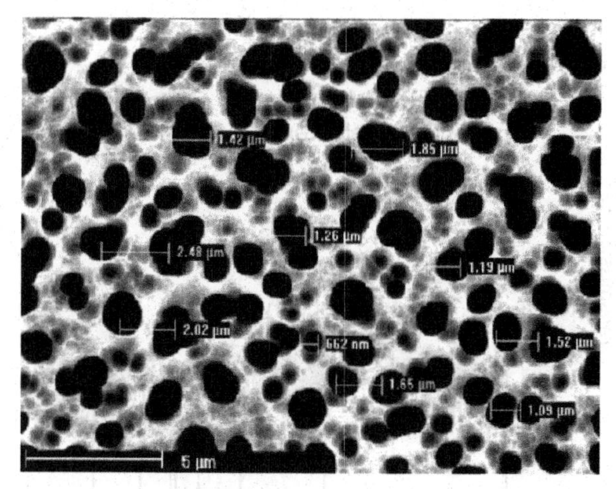

Figure 10.4. SEM image of random pores with neither long nor short range order used for demonstrating retroreflection suppression (after [2])

10.3 Experimental Demonstration

For optical demonstration of the retroreflection suppression an MPSi membrane with parameters suitable for retroreflection suppression in the UV spectral range was made. The fabrication of the random MPSi array is essentially similar to that reviewed in previous chapters and will not be repeated here. The SEM image of the top surface of MPSi layer used for optical testing is given in Figure 10.4.

For the demonstration of the retroreflection suppression the fabricated uncoated MPSi membrane shown in Figure 10.4 was mounted on the surface of an Al mirror and the normal incidence reflection from such a structure was recorded in a spectrometer within the "far" field of 20 cm next to the Al mirror/MPSI membrane surface. The result of the measurement is presented in Figure 10.5. One can see that even at 20 cm the retroreflection was suppressed by 5 orders of magnitude. For a comparison, the near field transmission (recorded by mounting the sample directly on a detector in the spectrometer) is also plotted in Figure 10.5. It should be noted that at realistic distances from the shielded FPA to an enemy detector of over 1 km (in most situations), more than 10 orders of magnitude of retroreflection suppression can be expected. One should note that the reflection was suppressed quite uniformly over the wide spectral range. As explained above, by coating the pore walls with metal the near field transmission can be increased even further.

In order to investigate the imaging properties of the detector array when used with MPSi retroreflection suppression plate, the MPSi membrane with larger pores (sufficiently large to provide good transmittance in visible spectral range) was mounted on the top of a commercially available CCD camera (as illustrated in Figure 10.6a. The image of a parking lot obtained through such a structure is presented in Figure 10.6b. The (inferior) resolution of the (very cheap) camera could essentially be maintained.

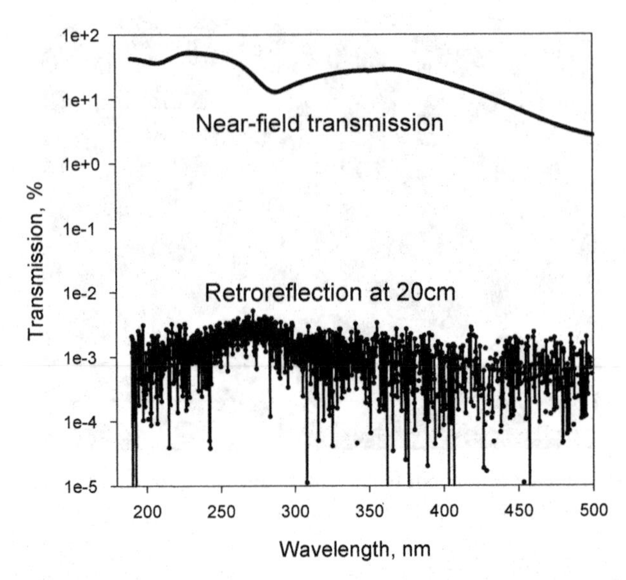

Figure 10.5. Measured spectral dependences of near field and far field retroreflection of the aluminum mirror/MPSi membrane structure. After [2]

Figure 10.6. a Photo of the camera with the MPSi membrane mounted on the top; **b** Image of a parking lot obtained with the camera and the MPSi membrane. After [2]

10.4 Conclusions

It is demonstrated that MPSi membranes can be used effectively for retroreflection suppression purposes, offering the following advantages:

1. High near-field efficiency.
2. Highly uniform scattering in the far-field.
3. Retroreflection will be suppressed over very wide wavelength range (from deep UV to far IR in one design).
4. Applicability to wavelengths range from deep to far IR.

5. The plate size is only limited by the Si wafer diameter, and wafers could be even “stitched” together.
6. Manufacturing is compatible with standard large scale Si technology.
7. The plate is very compact and lightweight (the weight of a 15 cm diameter plate would not exceed several grams).
8. No major changes of existing designs of the LWIR FPA mechanical holder would be needed.

In addition to military applications such a plate can be used in other applications suffering from stray light such as light such as microscopes and spectrometers. However, application potential of such components is still have to be realized. While no principal difficulties is expected in development of porous-semiconductor retroreflection plates, it is yet to be shown that the market potential can substantiate the necessary investments in product development.

10.5 References

[1] Coleman LA, (1994) Infrared Sensors: The Eyes of the Digital Battlefield. Military and Aerospace Electronics, 29.

[2] Kochergin V, Föll H, (2006) Novel optical elements made from porous silicon. Review Materials Science and Engineering R, 52:93–140.

[3] George N, Schertler DJ, (2001) Optical system for diffusing light. U.S. Patent 6,259,561.

[4] Klaus JW, Ferro SJ, George SM, (2000) Atomic Layer Deposition of Tungsten Nitride Films Using Sequential Surface Reactions. J. Electrochem. Soc., 147, 1175–1181.

[5] Hejjo Al Rifai M, Christophersen M, Ottow S, Carstensen J, Föll H, (2000) Potential, temperature and doping dependence for macropore formation on n-Si with backside illumination. J. of Porous Materials 7, 33–36.

[6] Lehmann V, Rönnebeck S, (2001) MEMS Techniques Applied to the Fabrication of Anti-Scatter Grids for X-Ray Imaging. in *Technical Digest of the 14th IEEE International Conference on Micro Electro Mechanical Systems*, 84, Interlaken.

11

Omnidirectional IR Filters

11.1 Introduction

Narrow-band pass, band-pass and band-blocking IR filters are used intensively in optical communications, imaging, spectroscopy, astronomy, and many other applications. Currently available filters are based on interference in multilayer stacks and have to be used with well-collimated beams and carefully aligned angles of incidence. The angular dependence of the pass band position for IR filters used in, for example, DWDM (Dense Wavelength Division Multiplexing) requires even tighter alignment (meaning complexity) than with UV filters discussed previously in this paper due to much more strict specifications. Omnidirectional IR filters, based on macroporous silicon technology, recently proposed in [1] and reviewed in [2], have the potential to solve this problem. This type of filters is based on macroporous silicon and utilizes the waveguide transmission mode discussed in Chapter 5. It will be analyzed here in more detail.

11.2 Theoretical Considerations

As was shown in Chapter 5, at wavelengths above ~ 1 μm an ordered MPSi layer can be considered as an array of waveguides. The upper limit of the waveguide transmission channel depends on the pore array period. Unlike the UV case, considered in previous chapters, waveguide losses cannot provide the transmission spectral characteristics needed for optical filtering applications. However, waveguides are known to exhibit the Bragg reflection phenomenon when their effective refractive indices are periodically modulated in the direction of the waveguide mode propagation. Modulation of effective refractive indices in silicon column waveguides in the MPSi layer is possible through modulation of the porosity (or, in other words, modulation of macropore diameters) of the MPSi layer with its depth (see Figure 11.1). The modulation of MPSi layer porosity will modify the spectral dependence of the transmission through each silicon island waveguide.

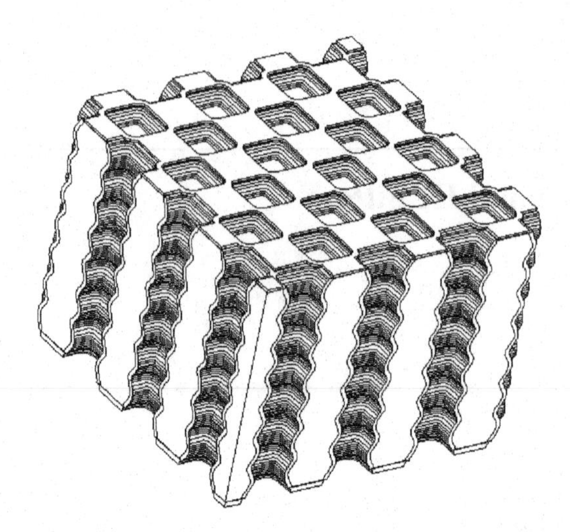

Figure 11.1. Schematic drawing of a massively parallel waveguide array with coherently modulated cross-sections. After [2]

The Bragg grating transfer function $T^{BG}(\lambda)$ depends on many parameters such as the length of the modulated waveguide part, the period (Λ) and the amplitude (σ) of the waveguide mode index (n^*) modulation that is related to the pore diameter modulation. It is also dependent upon the particular structure of the Bragg grating, i.e. whether it is uniform and periodic along it's length, or some phase-shifts (cavities in another terminology) are inserted [3]. Different permissible structures of Bragg gratings give great freedom in MPS-based filter design. In general, the quantity $\sigma l'/\lambda_B$ (where $\lambda_B = \Lambda/2\pi n^*$ is the Bragg wavelength, and l' is physical length of the grating element) should be sufficiently large in order to obtain good rejection or reflection levels.

In order to achieve the omnidirectionality of such filters, the spectral filtering process should be independent of the coupling/outcoupling processes (as with UV filters and polarizers discussed in the previous section). In other words, an array of *mutually decoupled* waveguides is required, so any transmission channel other than through the silicon columns waveguide modes needs to be suppressed. In Chapter 5 it was theoretically shown that although it is possible to suppress the cross-coupling between neighboring silicon columns by coating the pore walls with a silicon dioxide layer with an appropriate thickness, metal filling or metal-coating of the pore walls is a more effective method to insure the omnidirectionality of such a filter.

It should be noted that the angular range of such filters is predicted to be much wider than that of UV filters and is estimated to be more than +/–40° [1]. To illustrate the expected advantages of the spectral filter based on this design with respect to common interference filters, calculations are presented in Figures 11.2 and 11.3. Figure 11.2a gives the calculated transmittance spectra through a seven-cavity narrow-band-pass multilayer dielectric filter for normally incidence and 10- and 15-deg. tilted plane-parallel beams. The wavelength shift of the pass-band position together with the degradation of the pass-band shape is rather pronounced.

In Figure 11.2b, the calculated normalized transmittance spectra through a multiple-cavity MPSi-array-based filter are presented for normal incident and 20° and 30° tilted plane-parallel beams. The transmittance in this case is presented in a normalized form since the maximum transmittance is defined by the particular MPSi layer structure. Figure 11.3a presents the calculated transmittance spectra through the narrow-band-pass multilayer dielectric filter of Figure 11.2a for normally incident beams with different divergences: plane-parallel beam (0 deg.-divergence angle) and Gaussian beams with 10 deg. and 20 deg. divergence angles. The degradation of both the pass-band shape and the out-of-band rejection that are common to multiple-cavity all-dielectric multilayer filters are demonstrated. In Figure 11.3b, the calculations for normalized transmittance spectra through the multiple-cavity MPSi filter of Figure 11.2b are presented for 0 deg., 20 deg., and 40 deg. divergent, normally incident Gaussian beams.

In addition to omnidirectionality, the MPSi structure should have a high level of transmittance within the pass bands of $T^{BG}(\lambda)$. Further increase in transmittance is possible by minimizing the porosity at both MPSi layer interfaces, i.e. providing pores with "bottle-necks". However, the electrochemistry of silicon sets limits on the minimum pore cross-sections obtainable at the MPSi layer interfaces.

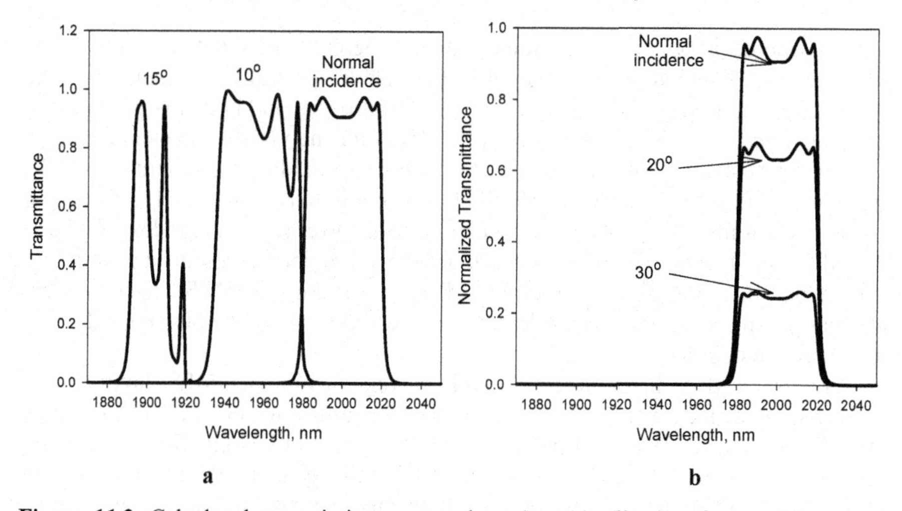

Figure 11.2. Calculated transmission spectra through a thin-film interference filter **a** and through an MPSi filter **b** for different angles (light incident at 0, 10 and 15 deg. form normal) of incidence (after [1])

The spectral filters for the IR range based on the MPSi layer structures discussed above promise significant advantages over dielectric multilayer-based filters. However, these designs cannot be directly transferred into the visible and near IR spectral ranges (400 nm to 1100 nm wavelengths) due to the absorption of the silicon, which would cause unreasonable propagation losses in silicon column waveguides. Nevertheless, visible and near IR spectral ranges are of great commercial importance and omnidirectional narrow-band-pass, band-pass, and band-blocking spectral filters are clearly needed for these wavelengths.

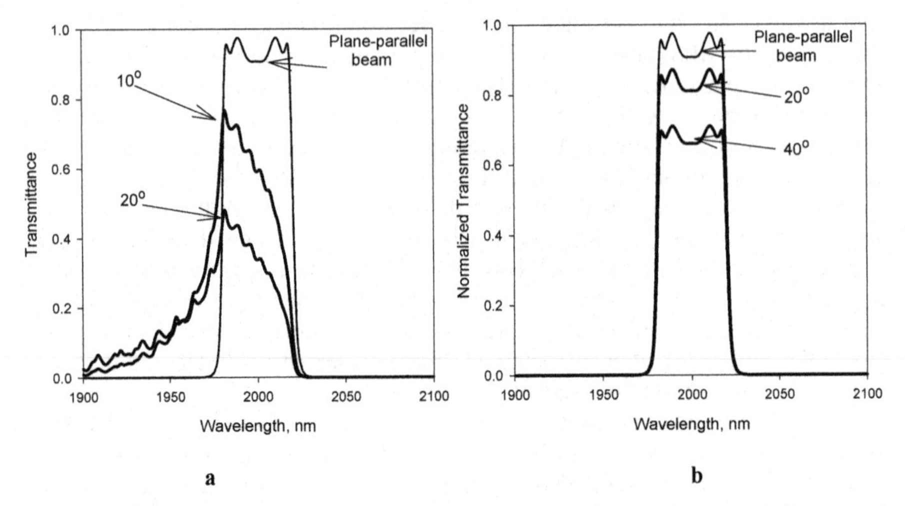

Figure 11.3. Calculated transmission spectra through a thin-film interference filter **a** and through an MPSi filter **b** for different divergences of Gaussian beams (after [1])

An ordered array of waveguides with a coherently modulated cross-section made of materials that are transparent in the visible spectral range (similar to the MPSi layer discussed above) would serve such purposes. In [1] it was proposed to use a *completely oxidized MPSi layer* (*COMPSi*) for the visible wavelength range. Lehmann has shown that it is possible to completely oxidize MPs membranes [4], and the considerations for MPSi IR filters can be directly transferred to COMPSi layers, except for some normalization caused by a lower refractive index of silicon dioxide than that of silicon. In particular, the reflection losses during coupling should be about 4% at each interface of a freestanding COMPSi layer and the acceptance angle of silicon-dioxide-based waveguides should be also considerably less than that of silicon.

As with an MPSi layer, silicon dioxide waveguides in a COMPSi layer cannot be considered independent of each other. In order to obtain visible-range, omnidirectional, narrow-band-pass, band-pass, or band-blocking filters based on the COMPSi layer, the structure of the COMPSi layer should be modified to suppress cross-coupling between neighbouring silicon dioxide waveguides. Coating the pore walls with metal layer can do this.

Another design of omnidirectional filters for visible wavelength range was suggested in [5]. The idea is to coat the walls of MPSi layer with modulated pore diameters with a transparent layer having a low refractive index, and then to fill the pores (with small openings in the center) with a transparent higher refractive index material. This can be done in principle by ALD techniques (or by combination of thermal oxidation and an ALD technique) discussed in more detail in relation to UV filters in Chapter 8.

Of course, the fabrication of such structures is more challenging than the fabrication of omnidirectional MPS structures for the IR range and the feasibility of such designs needs still to be proven experimentally.

11.3 Fabrication

The general formation of MPSi was reviewed in detail in Chapter 2. In this section the discussion will be focused on some peculiarities of the fabrication process originating from the particular symmetries of pore arrangements required by filter design.

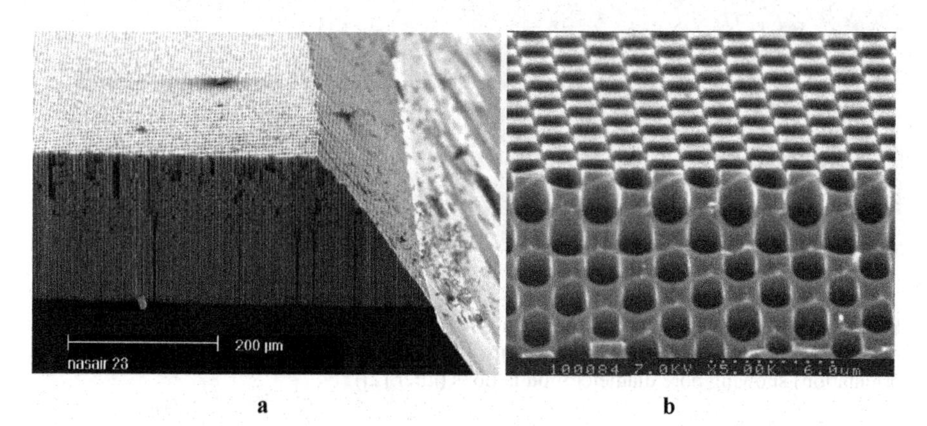

Figure 11.4. SEM images of an MPSi layer having the structure of Figure 11.1: Tilted views at different magnifications and orientation. The top surface of the MPSi layer corresponds to the (100) crystallographic orientation, while the front surface in **a** is (010) oriented while in **b** is (111) orientated (after [2])

The peculiar ("chess-board") arrangement of silicon islands of Figure 11.1 was achieved with the help of common photolithographic equipment and anisotropic etching of Si in a KOH solution, providing well developed inverted pyramids as nuclei. The pore diameter modulation was obtained by periodic variations of the current density during the electrochemical etching of p-doped (100)-oriented silicon wafers in organic electrolyte. Figures 11.4 and 11.5 show SEM images of different cross-sections of MPS arrays having modulated macropore diameters and a pore geometry as shown in Figure 11.1. While a fair matching to the initial design goals was achieved, up to now the proper pore modulation parameters was not achieved. This prevented an experimental proof of the feasibility of the discussed filter design. However, much progress in pore diameter modulation etching has been achieved in the meantime by the Halle group [6,7], and it is believed that further optimization of the electrochemical etching process will enable the fabrication of functional omnidirectional IR filter structure using MPSi.

MPSi arrays with advanced symmetries as discussed in Chapter 5 were fabricated as well. The SEM images of an array with an "advanced hexagonal" symmetry is shown in Figure 11.6; the pattern symmetry is preserved down to at least 200 μm. While so far such arrays were successfully fabricated only on n-doped (100)-oriented silicon substrates with back-side illumination, it is believed that careful optimization of electrochemical etching condition can permit fabrication of similar MPSi arrays on p-doped (100)-oriented silicon substrates as well. No modulation of the macropore diameters in advanced symmetry arrays were attempted so far at all.

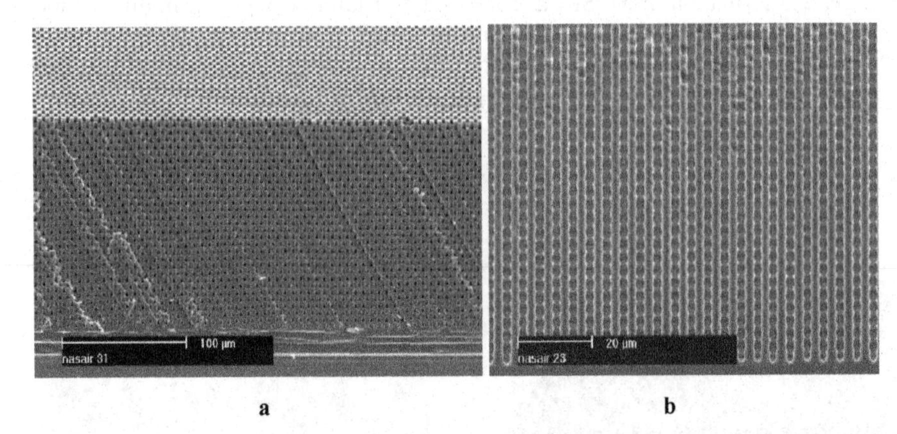

Figure 11.5. SEM images of the same MPSi layer as in Figure 11.4: **a** different magnification of the same view as in Figure 11.4b; **b** Magnified cleaved face (across (010) orientation) showing pore diameter modulations (after [2])

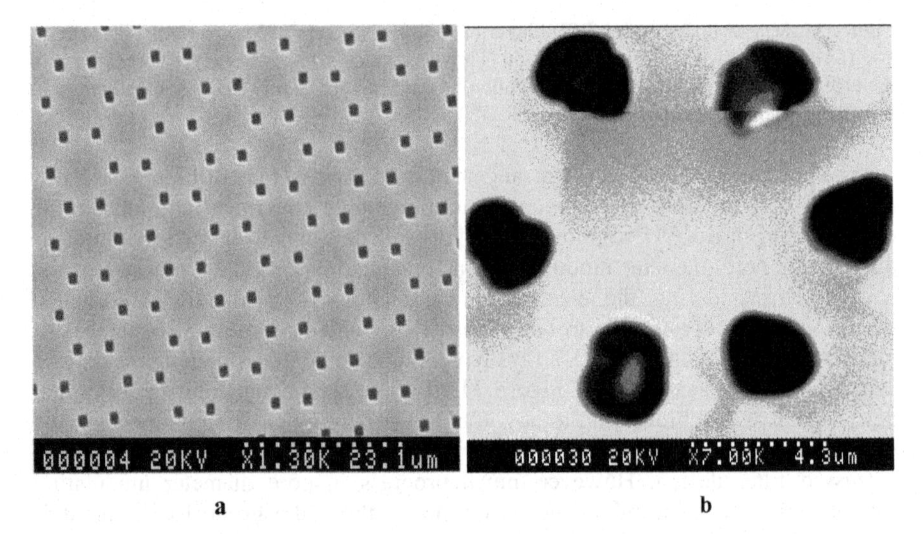

Figure 11.6. SEM images of the advanced hexagonal symmetry MPSi array. **a** Front surface. **b** Back surface (200 µm deep), opened by chemical etching (after [2])

Some experiments with metal filling of the pores by electro-plating were performed as a proof of principle as well. Plating was performed on unoxidized MPSi layers right after the anodization. A chamber similar to that used for silicon anodization was used for plating. It was found to be necessary to insure that the current is flowing only through the MPSi layer. Otherwise plating was mostly observed in the areas surrounding the MPSi layer.

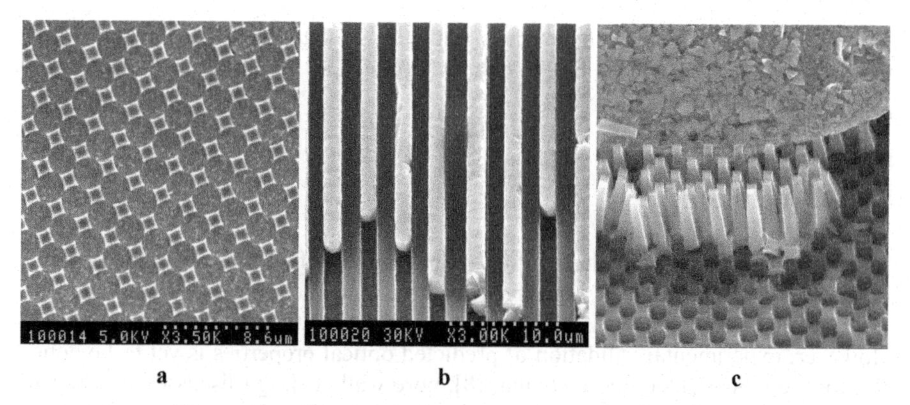

Figure 11.7. SEM images of the cupper-plated MPSi arrays

In such an electro-plating geometry the plating process works essentially inverse to the anodization process – i.e. sharp pore tip "focuses" current lines at the bottom of the pores so the plating initiated exactly at the bottom, not on the pore walls, and the plated metal continuously fills the pores from the bottom and up. Such a plating provided very good quality, voidless filling of the pores. The copper and permaloy (Fe/Ni alloy) were plated so far, although it is believed that any other metal can be equally plated into the pores as well. For copper plating, Cubath SC makeup provided by Enthone OMI was used as an electrolyte. The current density was 10 mA/cm^2 and was set in a DC mode. Figure 11.7 gives SEM images of the electroplated MPSi layers with copper completely filling the pores. Figure 11.7a shows the top view of the MPSi layer that was electroplated for ~1 hour, so the copper almost completely filled the pore, but unplated silicon "islands" were left on the filter "surface". Figure 11.7b gives an SEM image of the cleaved edge of the MPSi layer. Good quality of the pore-filling copper is clearly demonstrated. The uneven filling of the pore was not due to the bad plating but rather occurred during the cleavage of the wafer – copper has quite poor adhesion to silicon and during cleavage part of the pore-filled copper was left on one side of the cleaved wafer while other part of pore-filled copper was left on another part of the cleaved wafer. It also should be noted that all the pores were filled uniformly. Figure 11.7c shows an SEM image of the MPSi layer surface near the cleavage for the MPSi layer that was severely over-plated (2 hours), so a thick layer of copper was formed over all the surface of the MPSi layer. It should be noted that cleaving such a wafer was difficult because of the malleability of copper, so multiple bending of the copper layer was required prior to separating the pieces. During this procedure, part of the copper was removed from the pores, as illustrated in the image. One can see that not only the copper columns were uniform, but also the quality of the copper in the columns was considerably better than that covering the MPSi layer from the top. This is another indication of the good-quality plating achieved.

Note also that the copper-plating process used here is suitable for plating only pure silicon MPSi layers attached to the non-porous part of the Si wafer and cannot be used with MPSi layers having pores coated by non-conductive materials or free-standing MPSi layers. For such layers, the electro-plating process must be modified.

The work is currently in progress with the goal to fabricate the functional omnidirectional IR MPSi filter and to prove the feasibility of theoretical predictions.

11.4 Conclusions

Macroporous silicon IR filters promise a number of previously unachievable characteristics such as, e.g. omnidirectionality. The theoretical understanding of principles of operations of MPSi UV filters is believed to be well developed [1,2] However, experimental validation of predicted optical properties is yet to be done. Recent progress in macropore etching [8], pore wall coating (discussed in Chapter 8) and electroplating (briefly reviewed in this chapter) make authors believe that demonstration of omnidirectional MPSi IR filters can be achieved in the near future. It is hard to predict at this point whether marketable products based on this technology will be made though due to the considerable complexity of fabrication process. Porous layers of different semiconductors (such as InP [9]) may provide a basis for similar type optical filters as well. It would be also interesting to evaluate the possibility of making active components (e.g., modulators) on this technology.

11.5 References

[1] Kochergin V, (2003) Omnidirectional Optical Filters. Kluwer Academic Publishers, Boston, ISBN 1-4020-7386-0.
[2] Kochergin V, Föll H, (2006) Novel optical elements made from porous silicon. Review Materials Science and Engineering R, 52:93–140.
[3] Othons A, Kalli K, (1999) Fiber Bragg Gratings: Fundamentals and Applications in Telecommunications and Sensing, Artech House.
[4] Lehmann V, (2005) Porous silicon matrix for chemical synthesis and chromatograpy. Phys. Stat. Sol. 202:1365–1368.
[5] Kochergin V, Swinehart PR, (2003)U.S. Patent Application 20040134879.
[6] Matthias S, Mueller F, Jamais C, Wehrspohn RB, Goesele U, (2004) Large-Area Three-Dimensional Structuring by Electrochemical Etching and Lithography. Adv. Mat. 16: 2166–2170.
[7] Matthias S, Mueller F, Schilling J, Goesele U, (2005) Pushing the limits of macroporous silicon etching. Appl. Phys. A 80: 1391–1396.
[8] Föll H, Christophersen M, Carstensen J, Hasse G, (2002) Formation and application of porous Si. Mat. Sci. Eng. R 39: 93–141.
[9] Föll H, Langa S, Carstensen J, Christophersen M, Tiginyanu IM, (2003) Review: Pores in III-V Semiconductors. Adv. Materials, 15:183–198.

12

Biochemical Sensors Based on Porous Silicon

12.1 Introduction

Very high surface area and high reactivity, especially for meso- and nanoporous silicon, caused significant interest in scientific community in porous silicon-based chemical and biological sensors during last two decades. The first proposed transduction mechanism for porous silicon-based bio(chemical) sensors was the photoluminescence emission modified by the attached molecules to be detected [1]. Since then, other properties of porous silicon layers (such as the capacitance, refractive index, absorption, etc.) were proposed in a wide variety of configurations. Different functionalization chemistries of porous silicon surface developed over the years resulted in demonstration of wide range of sensors, while engineering the optical and/or electronic properties of porous silicon layers allowed a convenient integration of the sensing elements into optical or electronic devices. Sensing applications of porous silicon have been thoroughly reviewed [2,3] and we will provide only a brief overview of the state of the art particularly in the area of optical bio-chemical sensors.

12.2 Functionalization of Porous Silicon

From the chemical standpoint, pretty much any porous silicon bio-chemical sensor operates as follows: 1st, the porous silicon surface is properly functionalized such as only certain chemical or biological species are bound to the surface (to ensure selectivity), and 2nd, the functionalized porous silicon is exposed to the environment and the attachment (binding) of these chemical or biological species is detected by monitoring porous silicon properties (such as luminescence characteristics, dielectric properties, etc). Hence, functionalization of porous silicon is of utmost importance for sensing applications.

The native porous silicon surface is silicon-hydride ($Si\text{-}H_x$) terminated, with surface silicon atoms capped with one to three hydrogen atoms. The native $Si\text{-}H_x$-terminated surface is metastable, as illustrated in Chapter 6 (see Figure 6.15 and related discussion). This calls for a stabilization of the porous silicon surface prior to the incorporation of recognition molecules. Several approaches have been suggested for the stabilization of porous silicon layers prior to functionalization with the most popular being oxidation of the porous silicon surface and utilization of the native chemical reactivity of freshly etched porous silicon surfaces (which is typical for nano- and meso-porous silicon).

Oxidation of porous silicon is typically improving the stability of the material significantly. To provide even greater stability, the oxidized silicon surface can be chemically modified with alkyl silanes to provide a dense monolayer that limits the accessibility of the underlying surface to solution. A thorough review of oxidation and reactions involving Si-O bond formation can be found in [4]. Several variations of this technique can be mentioned in this regard: utilization of alkoxy- and chlorosilane chemistry [5,6], hydrosilylation of alkenes and alkynes induced by Lewis acids or white light [7,8], thermal alkyne hydrosilylation [9] to name a few. Silanization of oxidized porous silicon was successfully used to fabricate biorecognition interfaces composed of DNA [10,11,43], antibodies [12,13,44], enzymes [14–17] and small molecules [18,19], to give a few examples.

As an alternatively to the oxidation approach, hydrosilylation of alkenes and alkynes is used for the formation of monolayers bound to silicon surfaces through Si-C bonds. Such an approach first has been shown to be capable of protecting flat silicon surfaces from oxidation and chemically demanding environments [20–22], and was later adapted for porous silicon as well [7,8,23,24]. Utilization of such chemistries is believed to be beneficial for applications where photoluminescence efficiency should be preserved (it degrades during oxidation). Also, the formation of Si-C bonds on the surface allows for complex organic molecules to be covalently and irreversibly immobilized on the porous silicon surfaces. The chemistry of these processes was thoroughly reviewed recently [25]. Such a functionalization was successfully pursued to form porous silicon biorecognition interfaces composed of DNA [26], peptides [27], small molecules [28] and hybrid lipid bilayer membranes [29].

Below we will briefly overview a few examples for the realization of porous silicon sensors utilizing different sensing mechanisms.

12.3 Porous Silicon Sensors Based on Luminescence Quenching

Photoluminescence emission from nano- and meso-porous silicon can be quenched upon exposure to a variety of solvent vapors and gases [30–33]. This gives a natural method to provide a sensitive gas sensor. Photoluminescence quenching at its origin in a way is similar to the enhanced IR absorption discussed in Chapter 6: binding of the gas or vapor molecule to a surface site of silicon nanocrystallite can introduce 1) a site for nonradiative recombination of excitons, 2) interfacial charge, 3) energy transfer, 4) modification of the effective dielectric constant of the porous silicon layer. Depending on the type of the molecule adsorbed one or more of the mechanisms listed above can be responsible for photoluminescence quenching.

The main challenge in photoluminescence-based porous silicon sensors is to provide the specificity of the sensor while minimizing the quenching effects due to surface functionalization, which generally heavily influences the emission intensity [34]: Conjugated groups such as phenylacetylenyl [35] or styrenyl [34] groups completely quench porous silicon photoluminescence, while isolated double bonds only quench 40–80% of the photoluminescence efficiency. Phenyl termination has little effect on photoluminescence [35,36].

For example, it was shown that mouse monoclonal antibodies while inducing little change in photoluminescence efficiency when physisorbed onto porous silicon [37], cause large changes in photoluminescence efficiency upon binding to human myoglobin. Detectivity and accuracy of such sensors was favorably compared with a standard ELISA (Enzyme-Linked Immunosorbent Assay) test, while temporal response of porous silicon sensor was found to be by an order of magnitude faster (15 minutes compared to at least 3 hours).

Glucoronidase enzymes were also immobilized by alkene hydrosilylation on porous silicon films and a multiple step covalent coupling scheme was employed. Exposure of porous silicon sample to enzyme p-nitro-phenyl-beta-D-glucuronide in solution from 25 μM to 250 μM was detected by reduction in the photoluminescence concurrently with absorbance measurements [16]. Selena Chan and co-workers increased the sophistication of photoluminescence detection by building a resonant microcavity where the typically broad photoluminescence spectrum of porous silicon can be engineered into an interference pattern of sharp peaks. Decreasing the line width of the spectral features allows higher sensitivity for detecting small changes to the photoluminescence efficiency. This scheme was used to detect short DNA oligomers, large viral DNA molecules [38,39] and to discriminate between gram positive and gram negative bacteria colonies.

Photoluminescence-based porous silicon sensors were also demonstrated to be capable of selectively detecting nitric oxide and nitrogen dioxide [40], Nitrobenzene, DNT and TNT Vapors [41], TNT and picric acid on surfaces and in seawater [42] and other substances.

12.4 Porous Silicon Sensors Based on Change of Refractive Index

The very large surface-area to volume ratio of nano- and meso-porous silicon films permits significant changes of the refractive index of porous silicon even at small coverage of pore walls with analyte molecules. This property was exploited in a number of sensing concepts.

For example, interferometric (Fabry-Perot) porous silicon sensors were proposed by Sailor and co-workers [43–45]. It was suggested that biosensors based on such a principle offer label-free analyte sensing. By oxidizing porous silicon thermally or through ozonolysis, alkoxysilane linkers were attached to the surface through standard coupling chemistry. When derivatized surfaces of porous silicon are exposed to a solution of the complementary protein or DNA strand, binding occurs, which modifies the refractive index or optical thickness of the porous silicon layer. This induces a shift in the Fabry-Perot fringes which can be monitored with either spectrally or angularly. The principle of operation of porous silicon Fabry-Perot sensor is illustrated in Figure 12.1: binding of molecules on pore walls caused the red-shift of the Fabry-Perot interference pattern in reflection spectrum.

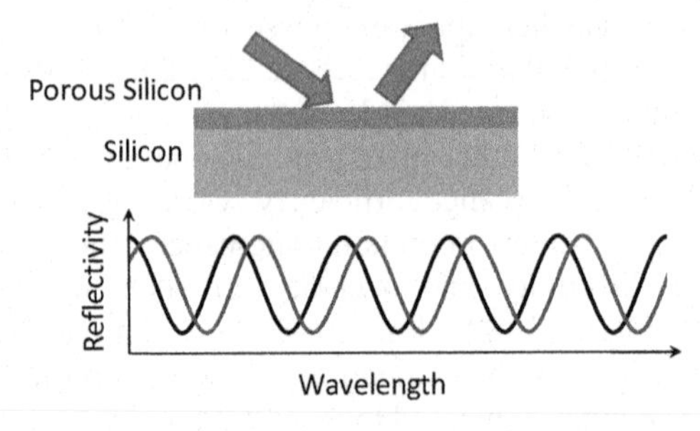

Figure 12.1. Schematic drawing illustrating operation of Fabry-Perot porous silicon sensor

Detection of DNA with pico- and even femtomolar concentrations were demonstrated with spectrally encoded Fabry-Perot porous silicon sensors. Similar sensors were employed to detect complimentary DNA proteins [46] and small molecules [47–49].

A highly selective HF sensor based on an oxidized porous silicon substrate formed through ozone oxidation was demonstrated, based on a similar technology [50]. In this case the change of the refractive index of the porous silicon layer was due to the dissolution of oxide on pore walls and the blue shift of the Fabry Perot fringes was used as a measured parameter.

In addition to a single porosity layer, double layer Fabry-Perot sensors were recently suggested to self-compensate sensors for signal drift [51].

The sensitivity of refractive index-based porous silicon sensors can be further enhanced by creating mesoporous silicon multilayer structure such as discussed in Chapter 6. In such a case the sensitivity and accuracy of the porous silicon sensors are enhanced due to much narrower spectral feature that experiences shift, as illustrated in Figure 12.2.

In contrast to multilobed spectra observed with single or double layer Fabry-Perot sensor, the porous silicon multilayer can be engineered to have one central cavity resonance with very narrow full width half maximum, as discussed in more detail in Chapter 6. Such an approach was employed by a number of groups for biochemical sensing. Rotiroty et al. demonstrated detection of pesticides in water samples [52] and De Stefano and co-workers demonstrated glutamine binding protein physisorbtion to mesoporous silicon for detection of glutamine in solution [53]. Using a strategy for pore enlargement to permit infiltration of relatively large biomolecules into mesoporous silicon, DeLouise and Miller demonstrated glutathioneS-transferase immobilization for processing solution phase substrates within the microcavity for identification of kinetic parameters for immobilized enzymes vs. solution phase systems with porous silicon resonant microcavity [54]. For even larger biomolecules, the use of macorporous silicon (with pores in the range of 50 to 200 nm) was suggested by Fauchet et al. [18]. Ouyang et al. detected 100 fmol on intimen extracellular domain binding to immobilized intimin binding domain (common recognition proteins responsible for E. coli pathogenicity) [55].

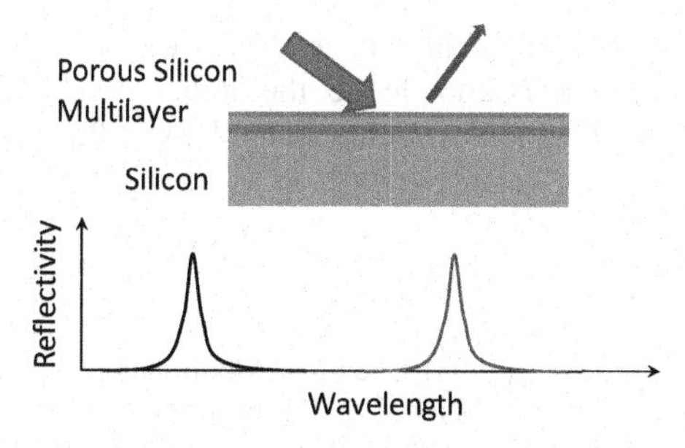

Figure 12.2. Schematic drawing illustrating operation of bragg reflectyor porous silicon sensor

Even "thinner" spectral reflection peaks than those obtained with typical porous silicon multilayer with more or less sharp porosity interfaces can be achieved with so-called porous silicon rugate filters (i.e., filters in which the refractive index is varied in a near-sinusoidal fashion). Such a porous silicon structure was also successfully used for sensing applications: Pavesi et al. demonstrated the usefulness of rugate porous silicon structures for detection of small biological interactions [56,57]. The Sailor research group has employed rugate filters for the development of "smart dust" microparticles [58]. Such porous silicon microparticles (fabricated by ultrasonic shredding of mesoporous silicon multilayers) were demonstrated for a number of biochemical sensing applications.

Other techniques to monitor the refractive index of porous silicon layer were demonstrated for sensing applications as well: Arwin and coworkers used ellipsometry to measure the change in index of refraction of a porous silicon matrix upon exposure to vapors from different liquids [59,60]. Gaseous concentrations of acetone as low as 12 ppm could be observed.

Refractive index detection by optical waveguide sensor was also realized with porous silicon technology for liquid sensing [61,62].

12.5 Conclusions

In the brief review given in this chapter we just barely touched the field of porous silicon-based sensors. We were focusing our discussion on optical sensors where porous silicon is playing a key portion by itself. It should be noted that a number of sensor realizations were suggested in which porous silicon served as a template. For example, polymer replicas [63] of mesoporous silicon multilayers were suggested for a number of biosensing applications. Filling the pores with metallic, [64,65] dielectric or semiconducting oxide [66,67] enzyme [68,69] or molecular receptor films [70] was realized to develop sensors capable of detecting penicillin, alkali metal ions, humidity, and hydrocarbons. Porous silicon, loaded with iron oxide nanoparticles was suggested for manipulating liquid droplets [71,72] and for performing localized heating [73]. Porous silicon was also used in mass spectrometry [74], flow-through reactors [75] and as a chemical release agent for in-vivo applications [76–78] to name a few.

While commercial bio-chemical sensors based on porous silicon are not yet available to the best of authors knowledge, the unique properties of porous silicon films promise commercialization of such a technology in the near future.

12.6 References

[1] Lauerhaas JM, Sailor MJ, (1993) Chemical Modification of the Photoluminescence Quenching of Porous Silicon. Science. 261:1567–1568.
[2] Sailor MJ, Lee EJ, (1997) Surface Chemistry of Luminescent Silicon Nanocrystallites. Adv. Mater. 9:783–793.
[3] Stewart MP, Buriak JM, (2000) Chemical and Biological Applications of Porous Silicon Technology Adv. Mater. 12:859–869.
[4] Song JH, Sailor MJ, (1999) Chemical modification of crystalline porous silicon surfaces. Comments Inorg. Chem. 21:69–84.
[5] Anderson RC, Muller RS, Tobias CW, (1993) Chemical Surface Modification of Porous Silicon. J. Electrochem. Soc. 149:1393–1396.
[6] Dubin VM, Vieillard C, Ozanam F, Chazalviel J-N, (1995) Preparation and Characterization of Surface-Modified Luminescent Porous Silicon. Phys. Status Solidi B, 190:47–52.
[7] Buriak JM, Allen MJ, (1998), Lewis Acid Mediated Functionalization of Porous Silicon with Substituted Alkenes and Alkynes. J. Am. Chem. Soc. 120:1339–1343.
[8] Buriak JM, Stewart MP, Geders TW, Allen MJ, Choi H-C, Smith J, Raftery D, Canham LT, (1999) Lewis Acid Mediated Hydrosilylation on Porous Silicon Surfaces. J. Am. Chem. Soc. 121:11491–11502.
[9] Bateman JE, Eagling RD, Worrall DR, Horrocks BR, Houlton A, (1998) Alkylation of Porous Silicon by Direct Reaction with Alkenes and Alkynes. Angew. Chem. Int. Ed. 37:2683–2685.
[10] Archer M, Christophersen M, Fauchet PM, (2004) Macroporous silicon electrical sensor for DNA hybridization detection. Biomedical Microdevices, 6:203–211.
[11] Steinem C, Janshoff A, Lin VSY, Voelcker NH, Reza Ghadiri M, (2004) DNA hybridization-enhanced porous silicon corrosion: mechanistic investigation and prospect for optical interferometric biosensing. Tetrahedron, 60:11259–11267.
[12] Bonanno LM, DeLouise LA, (2007) Steric Crowding Effects on Target Detection in an Affinity Biosensor. Langmuir 23:5817–5823.
[13] Schwartz MP, Alvarez SD, Sailor MJ, (2007) Porous SiO_2 Interferometric Biosensor for Quantitative Determination of Protein Interactions: Binding of Protein A to Immunoglobulin Derived from Different Species. Anal. Chem. 79:327–334.
[14] Chaudhari PS, Gokarna A, Kulkarni M, Karve MS, Bhoraskar SV, (2005) Porous silicon as an entrapping matrix for the immobilization of urease. Sensors and Actuators B, B107:258–263.
[15] Letant SE, Hart BR, Kane SR, Hadi MZ, Shields SJ, Reynolds JG, (2004) Enzyme Immobilization on Porous Silicon Surfaces. Adv. Mater. 16:689–693.
[16] Letant SE, Kane SR, Hart BR, Hadi MZ, Cheng T-C, Rastogi VK, Reynolds JG, (2005) Chem. Commun. 7:851–853.
[17] Reddy RPK, Chadha A, Bhattacharya E, (2001) Porous silicon based potentiometric triglyceride biosensor. Biosensors & Bioelectronics. 16:313–317.
[18] Ouyang H, Christophersen M, Viard R, Miller BL, Fauchet PM, (2005) Macroporous silicon microcavities for macromolecule detection. Advanced Functional Materials. 15:1851–1859.

[19] Chan S, Horner SR, Fauchet PM, Miller BL, (2001) identification of gram negative bacteria using nanoscale silicon microcavities. J. of American Chem. Soc. 123:11797–11798.
[20] Sieval AB, Demirel AL, Nissink JWM, Linford MR, van der Maas JH, de Jeu WH, Zuilhof H, Sudholter EJR, (1998) Highly Stable Si-C Linked Functionalized Monolayers on the Silicon (100) Surface. Langmuir. 14:1759–1768.
[21] Linford MR, Fenter P, Eisenberger PM, Chidsey CED, (1995) Alkyl Monolayers on Silicon Prepared from 1-Alkenes and Hydrogen-Terminated Silicon J. Am. Chem. Soc. 117:3145–3155.
[22] Bansal A, Li X, Lauermann I, Lewis NS, Yi SI, Weinberg WH, (1996) Alkylation of Si Surfaces Using a Two-Step Halogenation/Grignard Route. J. Am. Chem. Soc. 118:7225–7226.
[23] Buriak JM, (1999) Silicon Carbon Bonds on Porous Silicon Surfaces. Adv. Mater. 11:265–267.
[24] Boukherroub R, Wojtyk JTL, Wayner DDM, Lockwood DJ, (2002) Thermal hydrosilylation of undecylenic acid with porous silicon. J. of the Electrochem. Soc. 149:H59-H63.
[25] Buriak JM, (1999) Organometallic Chemistry on Silicon Surfaces: Formation of Monolayers Bound Through Si-C Bonds. Chem. Commun. 1051–1060.
[26] Lie LH, Patole SN, Pike AR, Ryder LC, Connolly BA, Ward AD, Tuite EM, Houlton A, Horrocks BR, (2004) Immobilization and synthesis of DNA on Si(111), nanocrystalline porous silicon and silicon nanoparticles. Faraday Discussions. 125:235–249.
[27] Kilian KA, Boecking T, Gaus K, Gal M, Gooding JJ, Peptide Modified Optical Filters for Detecting Protease Activity. ACS Nano.
[28] Hart BR, Letant SE, Kane SR, Hadi MZ, Shields SJ, Reynolds JG, (2003) New Method for Attachment of Biomolecules to Porous Silicon. Chem. Commun. 3:322–323.
[29] Kilian KA, Boecking T, Gaus K, King-lacroix J, Gal M, Gooding JJ, (2007) Hybrid lipid bilayers in nanostructured silicon: a biomimetic mesoporous scaffold for optical detection of cholera toxin. Chem. Commun. 1936–1938.
[30] Di Francia G, La Ferrara V, Quercia L, Faglia G, (2000) Sensitivity of Porous Silicon Photoluminescence to Low Concentrations of CH4 and CO. J. Porous Mater. 7:287–290.
[31] Song JH, Sailor MJ, (1997) Quenching of Photoluminescence from Porous Silicon by Aromatic Molecules. J. Am. Chem. Soc. 119:7381–7385.
[32] Kelly MT, Bocarsly AB, (1998) Coord. Chem. Rev. 171:251.
[33] Lauerhaas JM, Credo GM, Heinrich JL, Sailor MJ, (1992) Reversible Luminescence Quenching of Porous Si by Solvents. J. Am. Chem. Soc. 114:1911–1912.
[34] Allen MJ, Buriak JM, (1999) Photoluminescence of Porous Silicon Surfaces Stabilized Through Lewis Acid Mediated Hydrosilylation. J. Lumin. 80:29–35.
[35] Song JH, Sailor MJ, (1998) Functionalization of Nanocrystalline Porous Silicon Surfaces with Aryllithium Reagents: Formation of Silicon-Carbon Bonds by Cleavage of Silicon- Silicon Bonds. J. Am. Chem. Soc. 120:2376–2381.
[36] Kim NY, Laibinis PE, (1998) Derivatization of Porous Silicon by Grignard Reagents at Room Temperature. J. Am. Chem. Soc. 120:4516–4517.
[37] Starodub VM, Fedorenko LL, Sisetskiy AP, Starodub NF, (1999) Control of myoglobin level in a solution by an immune sensor based on the photoluminescence of porous silicon. Sens. Actuators B 58:409–414.
[38] Chan S, Fauchet PM, Li Y, Rothberg LJ, Miller BL, (2000) Porous silicon microcavities for biosensing applications. Physica Status Solidi A. 183:541–546.

[39] Chan S, Li Y, Rothberg LJ, Miller BL, Fauchet PM, (2001) Nanoscale silicon microcavities for biosensing. Materials Science & Engineering C. C15:277–282.
[40] Harper J, Sailor MJ, (1996) Detection of Nitric Oxide and Nitrogen Dioxide with Photoluminescent Porous Silicon. Anal. Chem., 68:3713–3717.
[41] Content S, Trogler WC, Sailor MJ, (2000) Detection of Nitrobenzene, DNT and TNT Vapors by Quenching of Porous Silicon Photoluminescence. Chem. Europ. J., 6:2205–2213.
[42] Sohn H, Calhoun RM, Sailor MJ., Trogler WC, (2001) Detection of TNT and Picric Acid on Surfaces and in Seawater Using Photoluminescent Polysiloles. Angew. Chem. Int. Ed., 40:2104–2105.
[43] Lin VS-Y, Motesharei K, Dancil K-PS, Sailor MJ, Ghadiri MR, (1997) A porous silicon-based optical interferometric biosensor. Science. 278:840–843.
[44] Dancil K-PS, Greiner DP, Sailor MJ, (1999) A Porous Silicon Optical Biosensor: Detection of Reversible Binding of IgG to a Protein A-Modified Surface J. Am. Chem. Soc. 121:7925–7930.
[45] Janshoff A, Dancil K-PS, Steinem C, Greiner DP, Lin VS-Y, Gurtner C, Motesharei K, Sailor MJ, Ghadiri MR, (1998) Macroporous p-type silicon Fabry-Perot layers. Fabrication, characterization, and applications in biosensing. J. Am. Chem. Soc. 120:12108–12116.
[46] Stefano LD, Rotiroti L, Rea I, Moretti L, Francia GD, Massera E, Lamberti A, Arcari P, Sanges C, Rendina I, (2006) Porous silicon-based optical biochips. J. of Optics A. 8:S540–544.
[47] D'Auria S, Champdore MD, Aurilia V, Parracino A, Staiano M, Vitale A, Rossi M, Rea H, Rotiroti L, Rossi A, Borini S, Rendina I, Stafano LD, (2006) J. of Physics: Condensed Matter. 18:S2019–S2028.
[48] Stefano LD, Rossi M, Staiano M, Mamone G, Parracino A, Rotiroti L, Rendina I, Rossi M, D'Auria S, (2006) Glutamine-binding protein from E. coli specifically binds a wheat gliadin peptide allowing the design of a new porous silicon-based optical biosensor. J. of Proteome Research. 5:1241–1245.
[49] Tinsley-Bown AM, Smith RG, Hayward S, Anderson MH, Koker LAG, Torrens R, Wilkinson A-S, Perkins EA, Squirrell DJ, Nicklin S, Hutchinson A, Simons AJ, Cox TI, (2005) Immunoassays in porous silicon interferometric biosensor comined with sensitive signal processing. Physica Status Solidi A. 202:1347–1356.
[50] Létant S, Sailor MJ, (2000) Detection of HF gas with a porous Si interferometer. Adv. Mater. 5:355–359.
[51] Pacholski C, Sartor M, Sailor MJ, Cunin F, Miskelly GM, (2005) biosensing using porous silicon double-layer interferometers: reflective interferometric fourier transform spectroscopy. J. of Americal Chem. Soc. 127:11636–11645.
[52] Rotiroti L, Stefano LD, Rendina I, Moretti L, Rossi AM, Piccolo A, (2005) Optical microsensors for pesticides identification based on porous silicon technology, Biosensors & Bioelectronics. 20:2136–2139.
[53] Stefano LD, Rea I, Rendina I, Rotiroti L, Rossi M, D'Auria S, (2006) Resonant cavity enhanced optical microsensor for molecular interactions based on porous silicon. Physica Status Solidi A. 203:886–891.
[54] DeLouise LA, Miller BL, (2005) Enzyme Immobilization in porous silicon: quantitative analysis of the kinetic parameters for glutathione-S-transferases. Anal. Chem. 77:1950–1956.
[55] Ouyang H, DeLouise LA, Miller BL, faucet PM, (2007) Label-free quantitative detection of protein using macroporous silicon photonic bandgap biosensors. Anal. Chem. 79:1502–1506.
[56] Lorenzo E, Oton Claudio J, Capuj nestor E, Ghulinyan M, Navarro-Urrios D, Gaburro Z, Pavesi L, (2005) Porous silicon-based rugate filters. Appl. Optics. 44:5415–5421.

[57] Lorenzo E, Oton Claudio J, Capuj nestor E, Ghulinyan M, Navarro-Urrios D, Gaburro Z, Pavesi L, (2005) Fabrication and optimization of rugate filters based on porous silicon. Physica Status Solidi A. 2:3227–3231.
[58] Sailor MJ, Link JR, (2005) Smart Dust: nanostructured devices in a grain of sand. Chem. Commun., 2005:1375–1383.
[59] Zangooie S, Bjorklund R, Arwin H, (1997) Vapor sensitivity of thin porous silicon layers. Sens. Actuators B. 43:168–174.
[60] van Noort D, Welin-Klinstron S, Arwin H, Zangooie S, Lundstrom I, Mandenius CF, (1998) Monitoring specific interaction of low molecular weight biomolecules on oxidized porous silicon using ellipsometry. Biosens. Bioelect. 13:439–449.
[61] Arrand HF, Benson TM, Loni A, Arens-Fischer R, Krüger M, Thönissen M, Lüth H, Kershaw S, (1998) Novel liquid sensor based on porous silicon optical waveguides. IEEE Photonics, Tech. Lett. 10:1467–1469.
[62] Arrand HF, Benson TM, Loni A, Arens-Fischer R, Krueger MG, Thoenissen M, Lueth H, Kershaw S, Vorozov NN, (1999) Solvent detection using porous silicon optical waveguides. J. Lumin. 80:119–123.
[63] Park JS, Meade SO, Segal E, Sailor MJ, (2007) Porous silicon-based polymer replicas formed by bead patterning. Phys. Status Solidi A-Appl. Mater. 204:1383–1387.
[64] Polishchuk V, Souteyrand E, Martin JR, Strikha VI, Skryshevsky VA, (1998) A study of hydrogen detection with palladium modified porous silicon. Anal. Chim. Acta. 375:205–210.
[65] Balucani M, Bondarenko V, Dolgyi L, La Monica S, Maiello G, Masini G, Yakovtseva V, Ferrari A, (1997) p75 Humidity Sensor Based on Partially Oxidized Porous Silicon. Solid State Phenom. 54:75–85.
[66] Angelucci R, Poggi A, Dori L, Cardinali GC, Parisini A, Tagliani A, Mariasaldi M, Cavani F, (1999) Permeated porous silicon for hydrocarbon sensor fabrication. Sens. Actuators. 74:95–99.
[67] Schöning MJ, Ronkel F, Crott M, Thust M, Schultze JW, Kordos P, Lüth H, (1997) Miniaturization of potentiometric sensors using porous silicon microtechnology. Electrochim. Acta. 42:3185–193.
[68] Bogue RW, (1997) Novel porous silicon biosensor. Biosens. Bioelectron. 12:xxvii–xxxix.
[69] Thust M, Schöning MJ, Frohnhoff S, Arens-Fischer R, Kordos P, Lüth H, (1996) Porous silicon as a substrate material for potentiometric biosensors. Meas. Sci. Technol. 7:26–29.
[70] Ben Ali M, Mlika R, Ben Ouada H, M'ghaïeth R, Maâref H, (1999) silicon as substrate for ion sensors. Sens. Actuators, 74:123–125.
[71] Dorvee JR, Derfus AM, Bhatia SN, Sailor MJ, (2004) Manipulation of liquid droplets using amphiphilic, magnetic 1-D photonic crystal chaperones. Nature Mater. 3:896–899.
[72] Thomas JC, Pacholski C, Sailor MJ, (2006) Delivery of Nanogram Payloads Using Magnetic Porous Silicon Microcarriers. Lab Chip 6:782–787.
[73] Park J-H, Derfus AM, Segal E, Vecchio KS, Bhatia SN, Sailor MJ, (2006) Local Heating of Discrete Droplets Using Magnetic Porous Silicon-Based Photonic Crystals. J. Am. Chem. Soc. 128:7938–7946.
[74] Wei J, Buriak JM, Siuzdak G, (1999) Desorption/Ionization Mass Spectrometry on Porous Silicon. Nature. 399:243–246.
[75] Herino R, Bomchil G, Barla K, Bertrand C, (1987) Porosity and pore size distribution of porous silicon. J. Electrochem. Soc. 134:1994–2000.
[76] Canham L, Reeves C, Newey J, Houlton M, Cox T, Buriak JM, Stewart MP, (1999) Derivatized Mesoporous Silicon With Dramatically Improved Stability in Simulated Human Blood Plasma. Adv. Mater. 11:1505–1509.

[77] Canham L, (1995) Bioactive Silicon Structure Fabrication Through Nanoetching Techniques. Adv. Mater. 7:1033–1037. Bowditch AP, Waters K, Gale H, Rice D, Scott EAM, Canham LT, Reeves CL, Loni A, Cox TI, (1999) In-Vivo Assessment of Tissue Compatibility and Calcification of Bulk and Porous Silicon. Mater. Res. Soc. Symp. Proc. 536:149–156.

[78] Li X, Coffer JL, Chen YD, Pinizzotto RF, Newey J, Canham LT, (1998) Transition Metal Complex-Doped Hydroxyapatite Layers on Porous Silicon. J. Am. Chem. Soc. 120:11706–11709.

Index